Studies in Computational Intelligence

Volume 747

Series editor

Janusz Kacprzyk, Polish Academy of Sciences, Warsaw, Poland
e-mail: kacprzyk@ibspan.waw.pl

The series "Studies in Computational Intelligence" (SCI) publishes new developments and advances in the various areas of computational intelligence—quickly and with a high quality. The intent is to cover the theory, applications, and design methods of computational intelligence, as embedded in the fields of engineering, computer science, physics and life sciences, as well as the methodologies behind them. The series contains monographs, lecture notes and edited volumes in computational intelligence spanning the areas of neural networks, connectionist systems, genetic algorithms, evolutionary computation, artificial intelligence, cellular automata, self-organizing systems, soft computing, fuzzy systems, and hybrid intelligent systems. Of particular value to both the contributors and the readership are the short publication timeframe and the world-wide distribution, which enable both wide and rapid dissemination of research output.

More information about this series at http://www.springer.com/series/7092

Jacek Grekow

From Content-based Music Emotion Recognition to Emotion Maps of Musical Pieces

 Springer

Jacek Grekow
Faculty of Computer Science
Bialystok University of Technology
Białystok
Poland

ISSN 1860-949X   ISSN 1860-9503   (electronic)
Studies in Computational Intelligence
ISBN 978-3-319-88968-9   ISBN 978-3-319-70609-2   (eBook)
https://doi.org/10.1007/978-3-319-70609-2

Printed on acid-free paper

This Springer imprint is published by Springer Nature
The registered company is Springer International Publishing AG
The registered company address is: Gewerbestrasse 11, 6330 Cham, Switzerland

To my great Family

Foreword

More and more multimedia systems use emotions as a factor in their practical applications. They try to recognize emotions in all forms of content in which humans are emotionally involved—in other words, in texts, music, or films. An example of such solutions is giant music libraries available via the Internet with music search systems using various criteria. Emotions turned out to be one of the more attractive and novel criteria for searching them. Just searching for compositions by title, composer and/or genre was not enough, and adding emotions as an option improved the attractiveness and usefulness of the search systems.

To be able to index music files in terms of emotions, they first have to be identified. Music emotion recognition (MER) is an interdisciplinary problem that affects such fields as signal processing, machine learning, psychology, music theory, and practice. MER requires researchers to be familiar with various distant fields, which on one hand is a complex task and on the other tremendously exciting.

This book presents particular issues that pertain to the creation of a music emotion detection system, such as: deciding on an emotion model, creating ground truth data, extracting and selecting features, and constructing a prediction model. The author focused on content-based MER, i.e., on analyzing the structures contained in music files and finding a relationship with emotions. He focused on examining audio as well as MIDI files and for each presented the relevant feature sets that describe them.

In this work, MER is presented as a classification and regression problem, which is closely connected with the selected emotion model: categorical and dimensional, respectively. Taking into consideration the way humans perceive emotions, the author decided to focus on perceived emotions in music and not felt emotions. During training data construction, he takes advantage of musicians' knowledge who expressed their opinions on the emotions they observed within musical excerpts. The author takes the reader through the specific stages of the emotion recognition system construction and presents the path from music files to the emotion maps created from them, which visualize emotion distribution over time. The book also puts forth practical applications of automatic emotion detection that were tested experimentally.

Emotions in music are very much connected with humans; we could say that they have a very human nature. They accompany man while composing, performing, or listening to music. Let us hope that computer systems evaluating emotions will not replace humans in creating them, but will only aid us in better understanding their relationship with music.

This book can be seen as a source of information for scientists, academicians, teachers, and students involved in constructing automatic emotion detection systems, as well as musicologists studying and analyzing musical compositions. Additionally, it can serve as an inspiration for other similar experiments conducted with the aim of analyzing emotion distribution in music.

Zbigniew W. Ras
Professor & KDD Lab Director
University of North Carolina Charlotte, USA
Professor, Polish-Japanese Academy of Information
Technology PJAIT
Warsaw, Poland

Preface

This book presents the most important issues with automated systems for music emotion recognition. These problems include emotion representation, annotation of music excerpts, feature extraction, and machine learning. The book concentrates on presenting content-based analysis of music files, which automatically analyzes the structures of a music file and annotates this file with the perceived emotions. Emotion detection in MIDI and audio files is presented.

In the experiments, the categorical and dimensional approaches were used, while for music file annotation, the knowledge and expertise of music experts with a university music education. The built automatic emotion detection systems enable the indexing and subsequent searching of music databases according to emotion. The obtained emotion maps of musical compositions provide new knowledge about the distribution of emotions in music and can be used to compare the distribution of emotions in different compositions as well as for emotional comparison of different interpretations of one composition.

Białystok, Poland
May 2017

Jacek Grekow

Contents

Chapter 1
Introduction

1.1 Motivation

Music and emotions have always been interwoven. Would we still listen to music if it didn't affect us emotionally? Would a composer create music without wanting to express emotions? Emotion is one of the main elements considered when people listen to music as well as when they create it.

Through the development of computer technology, particularly machine learning and content analysis, automatic emotion detection in music files has become possible. Once taught to recognize emotions, computers can exceed human capabilities in the quantity and accuracy of performed analyses of compositions. More and more frequently, systems that search Internet music databases have been adding the select an emotion option to the basic search parameters, which include such things as title, composer, genre, etc.

As a professional musician, I have always been fascinated with expressing emotions through music. Also, analysis of musical compositions taking emotions into account provides us with interesting new insights into their construction. How did, Beethoven, for example, shape the emotions of his compositions so that they are now considered masterpieces? How do the compositions of one composer differ emotionally from another? Why do some compositions affect us with a whole range of emotions while others only one? Can the way an emotion is shaped over time in a musical composition be seen and visualized? These are the questions I tried to answer in this work.

The aim of this book is to present the stages of building automated systems for music emotion recognition. This includes conducting experiments on various music file formats and using different approaches, in the direction of creating emotion maps of musical pieces. Another objective is to indicate some uses for the obtained emotion maps, in the form of systems detecting patterns in the course of emotions or systems for comparing musical pieces, taking into account the shaping of the emotions.

© Springer International Publishing AG 2018
J. Grekow, *From Content-Based Music Emotion Recognition to Emotion Maps of Musical Pieces*, Studies in Computational Intelligence 747,
https://doi.org/10.1007/978-3-319-70609-2_1

This book presents the particular stages of my research on emotion detection in music. At first, I studied emotions in MIDI files using the categorical approach, which was connected with creating my own MIDI features for detecting emotions. Then, I conducted experiments on recognizing emotion classes in audio files with features extracted using audio analysis tools tailored to Music Information Retrieval. The next stage was applying the dimensional approach to studying audio files and the creation of emotion maps on the arousal-valence emotion plane, which visualized the emotional structure of musical pieces over time. The results of the last stages introduce new, and nowhere before presented by other authors, research on comparing different performances of the same composition using emotion tracking, and finding performances that are more and less similar. The applications of the presented emotion maps of music files can vary widely, and this work does not exhaust them all, but just initiates them.

1.2   Organization of This Book

This book is divided into three parts. Part I focuses on representations of emotions in music as well the process of creating music data sets. The content presented in this part is intensively used in the remaining two parts, devoted to emotion detection in MIDI files in Part II and emotion detection in audio files in Part III.

In Chap. 2, I explain two popular approaches used do describe emotions, categorical and dimensional. Different models based on a discrete number of classes as well as models specifying emotion type using an axis on the emotional space are presented. The selected emotion models discussed here, which include four basic emotions— happy, angry, sad and relaxed—in the categorical approach and Russell's model in the dimensional, were then used in later experiments.

In Chap. 3, I present the process of creating ground truth data for emotion detection in MIDI and audio files. The process of music file annotation by music experts with a university music education is described. The collected ground truth is used in the remaining chapters.

Chapter 4 opens Part II, which focuses on emotion detection in MIDI files. This chapter presents a set of features extracted from MIDI files, assembled into four groups: rhythm, harmony, harmony-rhythm, and dynamic. It also introduces feature calculation methods and their potential to individually discriminate between emotion categories.

Chapter 5 puts forward emotion detection in classical music pieces in MIDI format; a hierarchical categorical model of emotions consisting of two levels was used. During feature selection, the most useful MIDI features were found for building a classifier recognizing four emotions.

Chapter 6 is the first chapter of Part III, which presents issues connected with emotion detection in audio files. This chapter focuses on some of the most relevant audio features for emotion detection in music files. These features were divided into three groups: timbre, rhythm, and tonal. Their meaning is presented and an analysis of the distribution of their values for audio excerpts labeled using four basic emotions was carried out.

In Chap. 7, I conducted experiments for detecting emotions in audio files using the categorical approach. I built classifiers for different combinations of feature sets, enabling distinguishing the most useful ones for individual emotions. The result of emotion tracking in music files is emotion maps, which visualize the distribution of four emotions over time.

Chapter 8 proposes a system for the analysis of emotions contained within radio broadcasts, which is a practical application of the categorical approach for emotion detection in audio files from the previous chapter. The obtained results provide a new, interesting view of the emotional content of radio station broadcasts.

Chapter 9 focuses on building emotion maps of musical compositions using the dimensional approach. Emotion recognition was treated as a regression problem, and a two-dimensional valence-arousal model was used to measure emotions. I also examined the influence of different audio feature sets—low-level, rhythm, tonal, and their combination—on arousal and valence prediction. On the basis of the created emotion maps, I propose selected features to analyze and compare musical compositions taking into account changes in arousal and valence over time.

Chapter 10 describes the final, most complex system for comparative analysis of musical performances by using emotion tracking. It is an example of applying the dimensional approach for emotion detection in audio files. Here, we discover which performances of the same composition are more similar and which are quite distant in terms of the shaping of arousal and valence over time.

Part I
Emotion in Music

Chapter 2
Representations of Emotions

2.1 Perception of Emotions

Emotions are a dominant element in music, and they are the reason people listen to music so often. We can ask ourselves the question: What can be the possible perception of emotions while listening to music? The psychologist Gabrielsson in his work [25] made a distinction between emotion perception into perceived and felt (induced) emotions. In the case of the former, we can perceive emotional expression in music without necessarily being affected ourselves; while in the latter, we have an actual emotional response to the music. Perceived emotion is the emotion recognized in the music, and induced emotion is the emotion experienced by the listener. Perceived and felt emotions are two alternatives that were the focus of psychology papers, such as those by Juslin and Laukka [48] and by Vuoskoski [108].

In our own work analyzing music recordings, we consider perceived emotion in music. During our experiments, experts with a university music education were asked to describe the emotions they perceived in music fragments and their opinions were then used to build a model of emotion prediction in music recordings.

2.2 Categorical Approach

Music emotion detection studies are mainly based on two popular approaches: categorical or dimensional. In the first, emotions are described with a discrete number of classes, affective adjectives, and in the second emotions are identified by axes. In the categorical approach, there are many concepts about class quantity and grouping methods. One of the first psychology papers that focused on finding and grouping terms pertaining to emotions was by Hevner [42]. As a result of the conducted experiment, there was a list of 66 adjectives arranged into eight groups distributed on a circle (Fig. 2.1). Adjectives inside a group are close to each other, the nature of

© Springer International Publishing AG 2018
J. Grekow, *From Content-Based Music Emotion Recognition to Emotion Maps of Musical Pieces*, Studies in Computational Intelligence 747,
https://doi.org/10.1007/978-3-319-70609-2_2

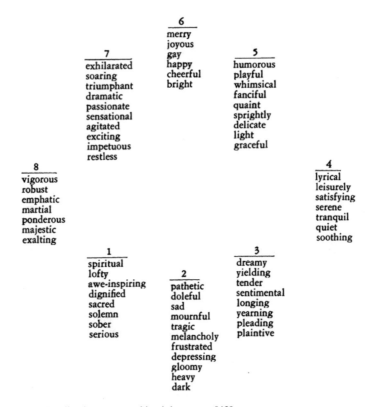

Fig. 2.1 Hevner's adjectives arranged in eight groups [42]

adjacent groups is evolving, and opposite groups on the circle are the furthest apart by emotion. Hevner's model was then modified by Farnsworth [23] and Schubert [97], who decreased the number of adjectives to 50 and 46, grouped them into nine groups.

Another interesting and important concept of finding the category of an emotion is the concept of basic emotion presented by Ekman [21, 22], which was developed for facial expression. Ekman describes features that enable differentiating basic emotions, which are:

- happiness,
- sadness,
- anger,
- fear,
- disgust,
- and surprise.

Ekman conducted experiments proving that facial expressions of basic emotions are cross-cultural. Johnson-Laird and Oatley [47] presented a somewhat smaller group of basic emotions: happiness, sadness, anger, fear, and disgust.

In the community of Music Information Retrieval Evaluation eXchange (MIREX) for automatic music mood classification, five mood clusters were used for song categorization [43]:

- Cluster 1 (passionate, rousing, confident, boisterous, rowdy);
- Cluster 2 (rollicking, cheerful, fun, sweet, amiable/good natured);
- Cluster 3 (literate, poignant, wistful, bittersweet, autumnal, brooding);
- Cluster 4 (humorous, silly, campy, quirky, whimsical, witty, wry);
- Cluster 5 (aggressive, fiery, tense/anxious, intense, volatile, visceral).

Hu et al. in [44] indicates, however, that the clusters might not be optimal and noticed some semantic overlap; similar findings were noted by Chen et al. [14]. The research carried out by Laurier et al. [55, 56] indicates deficiencies in this categorization, for example: experiments found that Cluster 1 and Cluster 5 are quite similar.

A popular emotion set used to categorize emotions in music turned out to be a collection consisting of 4 classes: happy, angry, sad, and relaxed. It corresponds to the four quarters of Russell's model [88], which were formed by dividing a plane by two perpendicular axes: arousal and valence. These values clearly define a point on the plane corresponding to a specific emotion and locate it on one of four quarters of Russell's model. The basic classes of emotions are assigned to the quarters as follows:

- happy—arousal high, valence high;
- angry—arousal high, valence low;
- sad—arousal low, valence low;
- relaxed—arousal low, valence high.

The selection of four categories of emotions also refers to the theory of basic emotions presented by Ekman [21]. The four categories are representatives of the main emotions from each of the quarters.

A significant disadvantage of the categorical approach is that the number of emotions and their shades perceived in music is much richer than the limited number of categories of emotions. The categorical approach has poorer resolution, by using the categories, we simplify the description of emotions in music, which facilitates understanding the character of the emotions and provides only a general overview of the emotions in music. One category contains an entire set of various shades of emotions. The smaller the number of groups in the categorical approach, the greater the simplification.

In this work, a set of four basic emotions: happy, angry, sad and relaxed, corresponding to the four quarters of Russell's model, were used for the analysis of music recordings using the categorical approach.

2.3 Dimensional Approach

In the dimensional approach, emotions are identified on the basis of their location in a space with a small number of emotional dimensions. In this way, the emotion of a song is represented as a point on an emotion space.

The two-dimensional circumplex model of emotion, which uses the two dimensions of arousal and valence, was presented by Russell in [88]. Arousal could be high or low and valence positive or negative (Fig. 2.2). In this model, all emotions can be understood as changing values of valence and arousal.

A variant of Russell's model is Thayer's model [103], in which the author suggested that two basic dimensions of describing emotions are two separate arousal dimensions: energetic arousal and tense arousal. In Thayer's model, valence could be explained as varying combinations of energetic arousal and tense arousal. Figure 2.3 is a visual presentation of the two models.

An example of a model where an emotion is described using three dimensions is Mehrabian and Russell's Pleasure-Arousal-Dominance (PAD) model [67], which was originally constructed to measure a person's emotional reaction to the environment. The three basic dimensions of emotions and their descriptions are: pleasure—positive and negative affective states; arousal—energy and stimulation level; dominance—a sense of control or freedom to act.

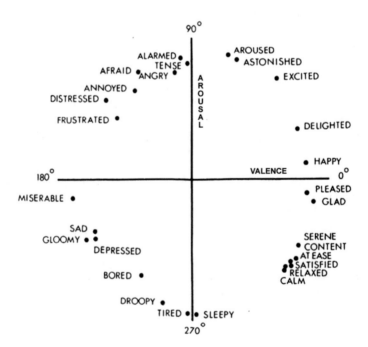

Fig. 2.2 Russell's circumplex model [88]

Fig. 2.3 Dimensional
models of emotions with
common basic emotion
categories overlaid. In
Russell's model, the axes are
indicated by a solid line; in
Thayer's model, the axes are
indicated by a dotted line
[20]

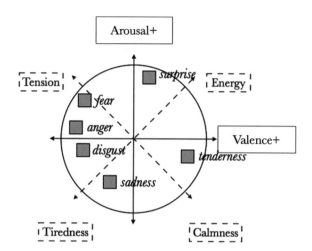

Experiments on automatic emotion prediction in music files using the PAD (Pleasure-Arousal-Dominance) and PA (Pleasure-Arousal) models were conducted by MacDorman et al. [64]. During file indexing, the authors noticed a significant correlation between the values of arousal and dominance. Ultimately, they decided to abandon the dominance dimension and use the Pleasure-Arousal model, because the results indicate that the dominance dimension was not informative for music.

A comparison of the discrete and dimensional models of emotion in music is presented by Eerola and Vuoskoski in [20], who used in their experiments five discrete emotions (anger, fear, sadness, happiness, and tenderness) and three bipolar dimensions (valence, energy arousal, and tension arousal). Linear mapping techniques between the discrete and dimensional models revealed a high correspondence along two central dimensions (valence and arousal). They concluded that the three dimensions could be reduced to two without significantly reducing the goodness of fit.

2.4 Summary

In our own work analyzing music recordings, we consider perceived emotion in music. Two approaches, categorical and dimensional, were used in emotion detection experiments. Using the categorical approach, a set of four basic emotions—happy, angry, sad and relaxed—were used. Using the dimensional approach, Russell's model—the most universal and least complicated to apply—was used. The four quarters of Russell's model correspond to four categories of emotions used in the categorical approach, which combines these two approaches to a certain degree. The categorical approach is more general and simplified in describing emotions, and the dimensional approach is more detailed and able to detect shades of emotions.

Chapter 3
Human Annotation

3.1 Introduction

Human annotation of music files is done in order to collect ground truth data, which will then be used to build an automatic emotion detection model. Annotation methods can be divided into two categories: expert-based and subject-based. Expert-based methods take advantage of music experts, musicians, people who work with music every day and most often play a musical instrument professionally. Music experts were used to determine music files in [39, 63, 104, 113]. Subject-based methods use people not connected professionally to music, non-musicians, and was employed in [2, 44, 51, 52, 64, 77, 105]. Subject-based methods also include a method of emotion tag collection directly from music websites such as Last.fm. This method was used in [54, 89, 116]. Due to the fact that survey respondents' answers can be somewhat subjective, the responses are often averaged.

In this paper, we used the expert-based method; we surveyed the opinions of five music experts, who are professionally involved with music every day, on the collected music samples. Each annotator annotated all music excerpts in the data set.

3.2 Length of a Musical Segment

An important element before carrying out annotation of music samples is to decide on the appropriate length of a musical fragment. The selected segment of a specific length will then undergo human annotation and audio features will be extracted. So what length should the indexed samples be or what is the shortest segment length necessary for an expert to be able to identify the emotion of a given fragment?

Many papers on automatic emotion detection have assumed a segment length of 20–30 s [44, 57, 63, 80, 104, 113]. These papers pertain to one static music emotion recognition, which assumes the emotion in a given segment does not change. Papers focusing on analyzing changes in emotions over time often use segments that are 1 s long [4, 51, 96, 119].

© Springer International Publishing AG 2018
J. Grekow, *From Content-Based Music Emotion Recognition to Emotion Maps of Musical Pieces*, Studies in Computational Intelligence 747,
https://doi.org/10.1007/978-3-319-70609-2_3

Selecting the appropriate segment length for emotion detection is quite important. On the one hand, a segment of a selected length is used during annotation by listeners—experts. On the other hand, it is used to extract audio features by a computer and to build an automatic emotion detection system. If to build training samples, we used a given length (i.e. a length of 6 s) of a musical segment, and on their basis we built emotion prediction models, prediction of new samples should also be of the same length, in our case 6 s.

The shorter a segment, the more detailed analysis of emotions is possible. Also, the shorter a segment, the more homogeneous the emotional content of the segment. On the other hand, a musical segment should not be too short during annotation since this will prevent a listener—expert from precisely identifying the emotion. Humans need time to determine the perceived emotion in music.

Bachorik et al. [5] investigated the length of time required for participants to initiate emotional responses to musical samples from a variety of genres by monitoring their real-time continuous ratings of emotional content and the arousal level of the music excerpts. On average, participants required 8 s of music before initiating emotional judgments.

Use of 6 s by Pampalk et al. [72] proved to be enough to build a system for content-based organization and visualization of music archives. From selected pieces of music in raw audio format, a geographic map was created where islands represented musical genres or styles.

Use of a segment shorter than 6 s hinders emotion detection by a listener [76, 111]. It enables differentiating two basic emotions, if a given fragment is happy or sad, but it prevents recognizing shades of emotions.

A segment with a duration of 6 s was used by Macdorman et al. [64] for automatic emotion prediction of song excerpts. At first 30-second segments were used, but due to the fact that pleasure and arousal typically change with musical progression, the segments were shortened to 6 s. The authors analyzed how pleasure and arousal ratings relate to the pitch, rhythm, and loudness of the song excerpts.

Xiao et al. [115] investigated the best segment duration for music mood analysis. Four versions of music data sets with a duration of clips of 4, 8, 16 and 32 s were tested. The results indicate that analyses of emotions in music should be based on shorter segments, no longer than 16 s, and the best performance was achieved by a segment length of 8 and 16 s.

In the report of the Emotion in Music task organized within the MediaEval benchmarking campaign, Aljanaki et al. [4] noticed problems with too short excerpts. The very short length of the annotated segments (0.5–1 s) allowed only capturing changes in dynamics and timbre. Simultaneously, it caused difficulty with capturing features pertaining to harmony and melody, which occur on a larger time scale.

In our experiment, the samples undergoing annotation were 6 s, which is the shortest possible length, determined experimentally, at which experts with a university music education could detect emotions for a given segment. A short segment

Table 3.1 Amount of examples of different genres of music

Genre	Amount of examples
Classical	67
Jazz	42
Blues	26
Country	50
Disco	27
Hip-hop	15
Metal	18
Pop	21
Reggae	22
Rock	36
All	324

ensures that the emotional homogeneity of a segment is much more probable. During the annotation, the experts sometimes provided their replies before 6 s were up, which suggests that a trained expert is able to identify an emotion before the end of 6 s.

3.3 Audio Music Data

3.3.1 Data Set

The data set that was annotated by the music experts consisted of 6-second fragments of different genres of music: classical, jazz, blues, country, disco, hip-hop, metal, pop, reggae, and rock. The tracks were all 22050 Hz, mono 16-bit audio files in .wav format. The training data were taken from the generally accessible data collection project MARSYAS.[1] The author selected samples and shortened them to the first 6 s, and as a result the data set consisted of 324 samples.

The amount of examples of different genres of music is presented in Table 3.1; the list of samples used in our experiments can be found on the web.[2]

---

[1] http://marsyas.info/downloads/datasets.html.
[2] http://aragorn.pb.bialystok.pl/~grekowj/HomePage/EmoDataSet.

3.3.2 Music Experts

Data annotation was done by five music experts with a university music education[3] (Expert 1–5), which included the author of this book. The experts, aged 28–48 years, are active musicians, playing on one of the following instruments: piano, clarinet, percussion, accordion, double bass. With their many years of experience, they are tied with various styles of music, such as: classical, jazz, blues, pop, rock, disco, punk.

3.3.3 Annotation Process

Before the annotation process, the music experts were introduced to Russell's model, with quarters corresponding to four basic emotions on the arousal and valence axes. Next, a 15-minute training was conducted during which the annotators listened to composition fragments and marked values on the arousal and valence axes. The meanings of arousal and valence were explained and differences between perceived emotion and felt emotion were discussed. In our experiment, the music experts' task was to identify perceived emotions only.

After this training and after teaching the experts about the applied terminology, they began annotating the musical segments. The annotation was carried out using a web application with a database specifically created for this purpose by the author of this book (Fig. 3.1). The application was built using Java Enterprise Edition, Java Server Faces, Server Glassfish, and the MySQL database. The web application enables access to the formulas indexing music files through a web browser, and the collected annotations were saved in the database.

Each annotator annotated all records in the data set, which had a positive effect on the quality of the received data [4]. The process was synchronized, as a sample was played for the whole group of experts simultaneously. After each sample was played, the experts had several seconds to make a decision, i.e. select values on the arousal and valence axes, and then the next composition was played.

The annotation process was repeated after the first round of annotation of all compositions was completed. The second annotation round enabled the experts to check and correct their responses. During the first round, it can be assumed that the music experts weren't completely trained in identifying valence and arousal values, while during the second round, they had the possibility of correcting their initial responses. The corrections weren't major, but nevertheless they occurred. Repeating the annotation process was beneficial because the experts were able to verify their responses; they had a second chance to make a decision and the possibility to change their response. During the second round of annotation, corrections mainly involved the first several samples from the first round. The closer to the end of the second

---

[3]We would like to thank the following music experts for indexing music files: Wojciech Bronakowski, Mateusz Bielski, Wojciech Mickiewicz, Jan Mlejnek.

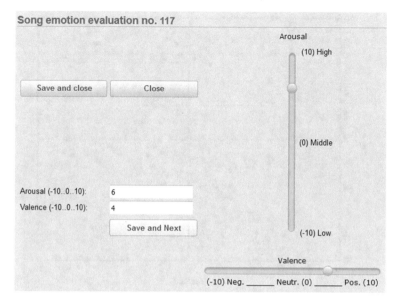

Fig. 3.1 View of the web application for labeling—survey for emotion labeling

round, the less corrections were being made. This indicates that the initial responses were not always correct, but the further into the indexing process the more correct the responses, which then did not need correction. There was a 30-minute break between the first and the second rounds of annotation. Despite the inconvenience, all annotators agreed with the validity of repeating the indexing process, since they were able to clarify their responses. Aljanaki et al. [4] also noted lower indexing quality at the beginning of their experiment; therefore, repeating the annotation process has a positive effect on the quality of the obtained data.

3.3.4 Results

As a result of the annotation by five music experts (Expert 1–5), we obtained data describing 324 fragments. Each fragment received five opinions, which is a total of 1620 of all annotations.

Figure 3.2 presents the 324 response, illustrated on the Arousal-Valence plane, from Expert 1. Each point represents one music excerpt, and its location on the plane is described by the arousal and valence values provided by Expert 1 for a given composition. As can be seen from the graph, responses can be found in all quarters of the A–V emotion model.

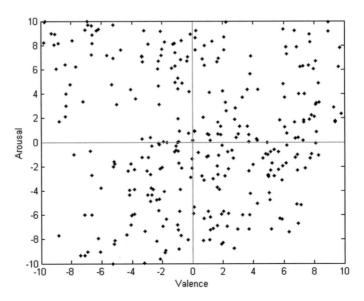

Fig. 3.2 Responses on the A–V emotion plane from Expert 1

Figure 3.3 presents the 324 responses from Experts 2–5. In all cases, there are responses in all quarters of the A–V emotion model, although the spread of point locations differ slightly.

Data collected from the five music experts were averaged; Fig. 3.4 presents the averaged responses for 324 compositions. Each point, with coordinates comprised of averaged values of arousal and valence, represents one music excerpt.

To check if in our music data valence and arousal dimensions are correlated, the Pearson correlation coefficient was calculated [13]. The obtained value $r = -0.03$ indicates that arousal and valence values are not correlated, and the music data are a good spread in the quarters on the A–V emotion plane. This is an important element according to the conclusions formulated by Aljanaki et al. in [4].

The values provided by the experts on the arousal and valence axes were in the range $[-10, 10]$. The mean of all collected values for arousal was: -0.16, and the mean for valence: 0.11. Both values are close to zero, which indicates a good distribution of samples on both sides of the valence and arousal axes. The mean standard deviation of the obtained responses from five experts on the arousal axis was 1.63, and on the valence axis 1.46. Both values are of a similar order and constitute about 8% of the entire range of values on the axes.

Considering the internal consistency of the collected data, Cronbachs α [15] for arousal and valence achieved high values of 0.94 and 0.86, respectively. From these values we can conclude that the agreement of the experts' opinions was greater for labeling arousal values than valence. Valence is usually more difficult to recognize, and here the experts' replies differed slightly more than in the case of arousal.

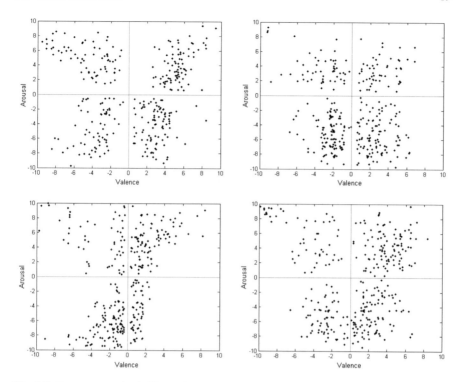

Fig. 3.3 Responses on the A–V emotion plane from Experts 2–5

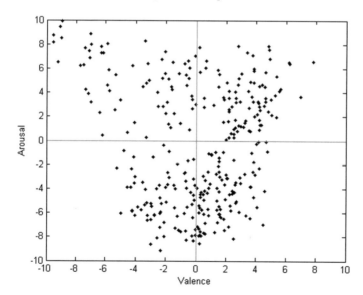

Fig. 3.4 Averaged values of responses from five music experts on the A–V emotion plane

Table 3.2 Amount of examples in quarters on the A–V emotion plane

Quarter abbreviation	Arousal-Valence	Basic emotion	Emotion label	Amount of examples
Q1	High-High	happy	e1	93
Q2	High-Low	angry	e2	70
Q3	Low-Low	sad	e3	80
Q4	Low-High	relaxed	e4	81

The amount of examples in the quarters on the A–V emotion plane is presented in Table 3.2. The arousal and valence values identify belonging to a given quarter of the model and emotion class simultaneously.

3.4   MIDI Music Data

3.4.1   Data Set

In this work, emotion detection experiments were conducted on audio files as well as MIDI files. For emotion detection experiments in MIDI files, we prepared a separate database with 83 compositions of classical music, which contains compositions by such eminent composers as:

- Franz Schubert (1797–1828),
- Ludwig van Beethoven (1770–1827),
- Felix Mendelssohn Bartholdy (1809–1847),
- Frédéric Chopin (1810–1849),
- Robert Schumann (1810–1856),
- Edvard Grieg (1843–1907),
- Isaac Albniz (1860–1909).

All compositions were piano-based; this way we rejected the aspect of studying the effect of various instruments on the perceived emotions. From the collected compositions, we extracted 350 six-second segments, an average of four fragments from each composition, which differed in tempo, volume, complexity, harmony, and dynamics.

The 350 music excerpts were annotated by five music experts with a university music education, people who have professional experience in playing and listening to music. To label emotions in MIDI files, we used a hierarchical model of emotions.

3.4.2 Hierarchical Emotion Model

The model we chose is based on Russell's circumplex model (Fig. 3.5). Following the example of this model, we created a hierarchical model of emotions consisting of two levels, L1 and L2.

The first level contains four categories of emotions that correspond to the four quarters of Russell's model (Table 3.3). In the first group (e1), pieces of music can be found that convey positive emotions and have a quite rapid tempo, are happy and arousing (excited, happy, pleased). In the second group (e2), the tempo of the pieces is fast, but the emotions are more negative, expressing annoying, angry, and nervous. In the third group (e3) are pieces that have a negative valence and low arousal, expressing sad, bored, and sleepy. In the last group (e4) are pieces that have low arousal and positive valence and express calm, peaceful, and relaxed.

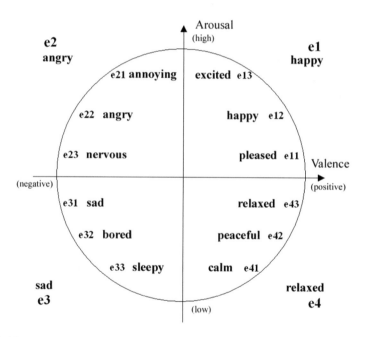

Fig. 3.5 Hierarchical model of emotions based on Russell's circumplex model

Table 3.3 Description of emotion categories in L1, the first level

Abbreviation	Description	Arousal-Valence
e1	happy	High-High
e2	angry	High-Low
e3	sad	Low-Low
e4	relaxed	Low-High

Table 3.4 Description of emotion categories in L2, the second level

Abbreviation	Description	Abbreviation	Description
e11	pleased	e31	sad
e12	happy	e32	bored
e13	excited	e33	sleepy
e21	annoying	e41	calm
e22	angry	e42	peaceful
e23	nervous	e43	relaxed

The second level is related to the first, and is made up of twelve sub-emotions, three emotions for each emotion contained in the first level (Table 3.4). In our hierarchical model of emotions, we have four categories in level L1 and twelve categories in level L2. The emotion categories in L1 are a generalization of the more detailed emotions in L2. Category names in L1 are also found in L2, for example, the entire group e1 in level L1 has been described by the adjective happy, and it includes the emotions excited, happy, and pleased in level L2.

3.4.3   Annotation Process

Six-second music samples were listened to and then labeled with one of the emotions of the second level (L2). Labeling with an emotion from the second level (L2) automatically indicated the parent emotion from the first level (L1). For example, if an expert selected emotion e13 (excited) from level L2, this meant they also selected emotion e1 (happy) from level L1. Thus, the samples were labeled with emotions from two levels of the hierarchical model. The short 6-second length of each segment ensured that the studied music fragments were relatively homogeneous emotionally, which allowed labeling a segment with one emotion. Each annotator annotated all records in the data set.

3.4.4   Results

The data collected from the five music experts were averaged by selecting an emotion that occurred the most often among the experts' responses. The amount of obtained examples labeled by emotions on the first level is presented in Table 3.5, and those labeled by emotions on the second level are presented in Table 3.6.

Considering the internal consistency of the collected data, Cronbachs α [15] obtained a value of 0.90 for data in level L1. Cronbachs α for collected data in level L2 obtained a value of 0.88, which means it was lower than for level L1. This

Table 3.5 Amount of MIDI examples labeled by emotions from the first level (L1)

Emotion abbreviation	Emotion	Amount of examples
e1	happy	89
e2	angry	105
e3	sad	79
e4	relaxed	77

Table 3.6 Amount of MIDI examples labeled by emotions from the second level (L2)

Emotion abbreviation	Emotion	Amount of examples
e11	pleased	39
e12	happy	36
e13	excited	14
e21	annoying	35
e22	angry	36
e23	nervous	34
e31	sad	37
e32	bored	23
e33	sleepy	19
e41	calm	19
e42	peaceful	15
e43	relaxed	43

is logical since we have more categories at level L2 as well as more differences in experts' opinions.

3.5 Summary

In this chapter, we presented two music data sets, audio and MIDI, that underwent annotation by music experts. We used specifically written web applications to collect data, which facilitated indexing musical compositions by many experts simultaneously. Each annotator annotated all records in the data set, and data collected from the music experts were averaged. The obtained music data are a good spread in the quarters of Russell's emotion plane.

The collected labeled audio music excerpts will serve as ground truth data during automatic emotion detection using the categorical (Chaps. 7 and 8) and dimensional (Chaps. 9 and 10) approaches. In the dimensional approach, we will use the collected arousal and valence values. In the categorical approach, we will use the four emotion classes corresponding to the four quarters of Russell's model: happy, angry, sad, and relaxed. The audio music excerpts labeled by four emotion classes will also be used for initial assessment of the usefulness of the designed audio features in Chap. 6 Sect. 6.3. The collected labeled music MIDI excerpts will serve as ground truth data during categorical hierarchical emotion detection in MIDI files in Chap. 5 as well as for initial assessment of the usefulness of the designed MIDI features in Chap. 4 Sect. 4.3.

Part II
Emotion Detection in MIDI Files

Chapter 4
MIDI Features

4.1 Introduction

In order to detect emotions in MIDI music files, we first need to perform an extraction of MIDI features that represent the musical content comprising the emotional content of music. In MIDI format [87] musical data is stored symbolically, containing higher-level abstractions, such as the pitch of each note, the instrument played, and the start and stop times of each note. In MIDI files, musical information is represented in a fundamentally different way than in audio files, which store an approximation of the sound waves produced by musical instruments. MIDI files store specific sounds, their beginnings and endings at indicated times, and thus we have direct access to the sounds played.

Symbolic data of musical information in MIDI files is much more familiar for musicians and musicologists than information contained in audio files. Musicians analyzing, discussing and playing music use the pitch and volume of sound, which is information that is already contained in MIDI files. The entirety of musicology knowledge pertaining to musical compositions, gathered over the centuries of the development of music, can be used to analyze files with symbolic data.

Features, which directly refer to rhythm, tempo, and harmony, extracted from MIDI files have a specific meaning for humans. It is easier to extract high-level musical features from MIDI files than from audio files.

Other formats that also store musical information as symbolic data include, for example, MusicXML [29] and Humdrum [45]. Of course, sheet music written by a composer is also a symbolic recording of music. We chose MIDI as the format for the studied music files due to its popularity and relatively large number of musical compositions available in this format. The presented features extracted from MIDI files could also be extracted from other formats that record symbolic data of musical information.

© Springer International Publishing AG 2018
J. Grekow, *From Content-Based Music Emotion Recognition to Emotion Maps of Musical Pieces*, Studies in Computational Intelligence 747,
https://doi.org/10.1007/978-3-319-70609-2_4

4.2   MIDI Features Tools

There are analysis tools tailored for Music Information Retrieval (MIR) that specialize in extracting music information from MIDI files, such as jSymbolic, MIDI Toolbox, or Melisma Music Analyzer. jSymbolic [66] software allows users to extract 160 features from a MIDI file; it is an open-source Java implementation and can save features in the ARFF format. In jSymbolic, a number of intermediate representations are prepared, including beat histograms, pitch histograms, histograms based on the instruments present, melodic interval histograms, vertical interval histograms, and chord type histograms. The extracted features can be divided into the following seven categories: instrumentation, texture, rhythm, dynamics, pitch statistics, melody, and chords.

The MIDI Toolbox [19] is a set of functions for analyzing and visualizing MIDI files in the MATLAB computing environment. It includes filtering functions, analytical tools relating to melodic contour, similarity, meter-finding, key-finding, and segmentation. The MIDI Toolbox contains cognitive analytic techniques that are suitable for context-dependent musical analysis.

The Melisma Music Analyzer [102] is a system for analyzing and extracting information from music. MIDI files can be used as the analyzer's input. The Melisma system consists of several modules, dedicated to such operations as: metrical analysis, grouping notes into phrases, separating the melody from the accompaniment lines, harmonic analysis, and key analysis.

4.3   Description of MIDI Features

We suggest the set of features presented below for emotion detection in MIDI files. The created features were divided into four groups: rhythm, harmony, harmony-rhythm, and dynamic. Harmony and harmony-rhythm are our own solutions that we created while studying the dissonance of simultaneously occurring sounds.

In addition to a description of the features, there are visualizations of their distributions in files labeled with four emotions (e1—happy, e2—angry, e3—sad, e4—relaxed). The collected music MIDI data labeled by music experts was presented in Chap. 3 Sect. 3.4.

The proposed features were used exclusively for describing classical music files, and the instruments used in the MIDI files were limited to the piano. This way the aspect of instrument type was eschewed, and rhythm, harmony, and dynamics were the main focus.

Each segment extracted from the MIDI file was described using a features vector calculated in MATLAB. We designed a collection of 63 features, which describe a given segment taking into account such musical elements as harmony, rhythm, and dynamics.

Fig. 4.1 Beat histogram for fragment of F. Chopin Etude Op. 10, No. 5

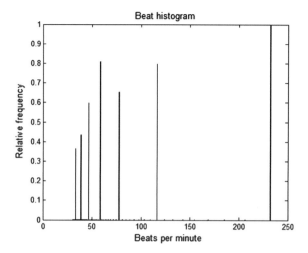

4.3.1 Rhythm Features

4.3.1.1 Beat Histogram

Rhythm features represent the rhythmic regularity in a given segment of music. Most of these features were obtained from the beat histogram, which was acquired from autocorrelation calculations [107].

$$autocorrelation[lag] = \frac{1}{N} \sum_{n=0}^{N-1} x[n]x[n - lag] \qquad (4.1)$$

where n is the input sample index (in MIDI ticks), N is the total number of MIDI ticks in a segment, and lag is the delay in MIDI ticks ($0 < lag < N$). The value of $x[n]$ is the velocity of Note On MIDI events.

The histogram was transformed so that each bin corresponded to a periodicity unit of beats per minute (BPM). The histogram values were normalized in relation to the highest value (magnitude) of the most frequent beat—the beat with the highest bar (Fig. 4.1).

The figures contain two beat histograms of two compositions by Frédéric Chopin. One of them is a segment from Etude Op.10 No.5 (Fig. 4.1) and the second from Preludium C minor Op.28, No.20 (Fig. 4.2). Etude is a relatively fast-paced composition with several dominating pulses. From the example of the image in Fig. 4.1, it is apparent that the First Strongest Rhythmic Pulse (SRP1) has a value of 240 BPM, the Second Strongest Rhythmic Pulse (SRP2) 60 BPM, and the Third Strongest Rhythmic Pulse (SRP3) 120 BPM. The contrast to this histogram is the histogram obtained for Prelude (Fig. 4.2). One main pulse (SRP1) of 40 BPM dominates, which is the result of the slow and uniform rhythm of this piece.

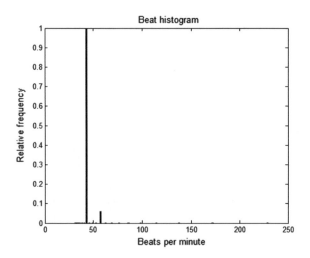

Fig. 4.2 Beat histogram for
fragment of F. Chopin
Prelude in C minor Op. 28,
No. 20

4.3.1.2   List of Rhythm Features

Rhythmic features extracted from a segment of music are presented in Table 4.1.
These features describe the strongest pulses in the piece (beats with the highest
magnitude in the beat histogram), the relations between them, and their quantity in
the beat histogram. The list also contains features describing the duration of notes
and the number of notes per second in a musical segment.

4.3.1.3   Selected Rhythm Features and Labeled Emotions

In order to visualize the usefulness of the MIDI rhythm features for emotion detection,
we present a spread of values of three selected features, for excerpts labeled by music
experts with four emotions (Figs. 4.3, 4.4 and 4.5).

When we observe the value distribution of the First Strongest Rhythmic Pulse
(SRP1) (Fig. 4.3), we notice that the spread of the median values for excerpts labeled
with emotions e1 (happy) and e2 (angry) have higher values than excerpts labeled
with emotions e3 (sad) and e4 (relaxed). SRP1 is connected with the composition's
tempo; a higher tempo results in higher SRP1 values. Also from a musical point of
view, musical fragments with sad and relaxed emotions will have a slower pulse,
which is confirmed by the presented lower SRP1 values in the box plot below.

From the box plot of Relatively Strong Pulses 30 (RSP30) (Fig. 4.4), we can see
that the music excerpts labeled with emotions from the higher quarters of Russell's
model of emotions (e1, e2) have higher median values than those from the lower
quarters (e3, e4). Also the spread (the space between the first and third quartiles) for
excerpts labeled as e2 (angry) is higher than for excerpts labeled with other emotions.
The lowest RSP30 values are for emotion e3 (sad). The higher RSP30 values for
excerpts labeled with emotions e1 (happy) and e2 (angry) indicate that the fragments

Table 4.1 Main rhythmic features

Feature	Description
First Strongest Rhythmic Pulse (SRP1)	The beats with the highest magnitude in the beat histogram
Second Strongest Rhythmic Pulse (SRP2)	
Third Strongest Rhythmic Pulse (SRP3)	
Ratios of the Strongest Rhythmic Pulses	The First Strongest Rhythmic Pulse divided by the Second Strongest Rhythmic Pulse: SRP1/SRP2
Ratios of the Magnitude of the Strongest Rhythmic Pulses	Magnitude of the SRP1 divided by magnitude of the SRP2: Magnitude(SRP1)/Magnitude(SRP1)
Relatively Strong Pulses 50 (RSP50)	The number of pulses with magnitude greater than 50% (30%, 10%) of magnitude of the SRP1
Relatively Strong Pulses 30 (RSP30)	
Relatively Strong Pulses 10 (RSP10)	
Note Density	Average number of notes per second
Average Note Duration	Average duration of notes in seconds
Standard Deviation of Note Duration	Standard deviation of note duration in seconds

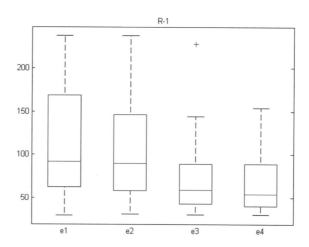

Fig. 4.3 Box plot of the first Strongest Rhythmic Pulse (SRP1) feature for MIDI data set labeled with four emotions e1–e4

have more varied strong pulses than excerpts labeled with emotions e3 (sad) and e4 (relaxed), which usually have less pulses competing with each other, which is completely logical from a musical point of view. From the high RSP30 values in fragments labeled as e2 (angry), we can conclude that a too great number of strong pulses elicits negative emotions in a listener. Extreme median values for emotions e2 and e3, very high for e2, very low for e3, and simultaneously the in-between values of segments with dominating emotions e1 and e4, suggest the usefulness of

Fig. 4.4 Box plot of the
Relatively Strong Pulses 30
(RSP30) feature for MIDI
data set labeled with four
emotions e1–e4

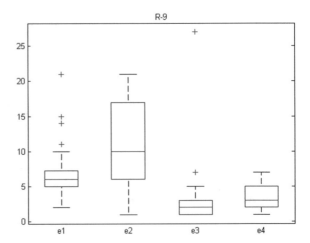

Fig. 4.5 Box plot of the
Average Note Duration
feature for MIDI data set
labeled with four emotions
e1–e4

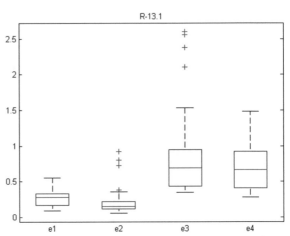

this feature to differentiate emotions on the valence axis, between the left and the right quarters of Russell's model. The usefulness of RSP30 was confirmed during the feature selection process while building classifiers (Chap. 5 Sect. 5.5.1.2).

When we observe the value distribution of the Average Note Duration feature (Fig. 4.5), we notice higher median values and a higher spread for excerpts labeled with emotions e3 (sad) and e4 (relaxed). The slower tempo of sad and relaxed music is connected with a longer duration of the played notes. Excerpts labeled with emotions e1 (happy) and e2 (angry) have clearly lower values. This confirms that the Average Note Duration feature could be useful for differentiating emotions on the arousal axis, between e1, e2 and e3, e4.

4.3.2 Harmony Features

Harmony, along with rhythm and dynamics, is one of the main elements of music upon which emotion in music is dependent. Harmony features reflect dissonance and consonance of harmony of sounds. They are based on my previous work, where I presented visualizations of harmony [30, 31, 33, 37].

To calculate the harmony parameters, we used the frequency ratio of simultaneously occurring sounds (Table 4.2). A given consonance (interval, chord, polyphone) comprises of simultaneously resonant sounds, the frequency ratio of which can be noted as follows:

$$N_{R1} : N_{R2} : \ldots : N_{Rk} \tag{4.2}$$

where k is the number of sounds comprising the consonance. N_{Ri} is taken from the just intonation tuning system, where the frequencies of the scale notes are related to one another by simple numeric ratios.

4.3.2.1 Chord Degree of Dissonance

From the frequency ratios, we calculated the *AkD* parameter, which mirrors the degree of dissonance in a single chord:

$$AkD = LCM(N_{R1}, N_{R2}, \ldots, N_{Rk}) \tag{4.3}$$

where k is the number of sounds in a given sample. In the case when $k = 1$, then $AkD = 1$. *LCM* means Least Common Multiple. The higher its value, the more dissonant the consonance; when the *AkD* value is lower, the consonance is more consonant—more pleasant for the ear.

From the sequence of consonance samples collected from a musical segment (Fig. 4.7), the array AkD_S can be defined as:

$$AkD_S = (AkD_1, AkD_2, \ldots, AkD_p) \tag{4.4}$$

where p is the number of samples collected from a given segment.

Table 4.2 Example consonance sound frequency ratios

Number of sounds	Musical notes	Consonance sound frequency ratios
2	$C_1 : G_1$	$2 : 3$
3	$C_1 : E_1 : G_1$	$4 : 5 : 6$
4	$C_1 : E_1 : G_1 : B\flat_1$	$25 : 30 : 36 : 45$

4.3.2.2   Mean of Frequency Ratio of Sounds in Chord

From the frequency ratios of simultaneously occurring sounds, we also calculated
the *AkM* parameter:

$$AkM = mean(N_{R1}, N_{R2}, \ldots, N_{Rk})$$ (4.5)

where k is the number of sounds in a given sample.

From the sequence of *AkM* from a musical segment, the array AkM_S can be defined
as:

$$AkM_S = (AkM_1, AkM_2, \ldots, AkM_p)$$ (4.6)

where p is the number of samples collected from a given segment.

4.3.2.3   Multiplication of Frequency Ratio of Sounds in Chord

From the frequency ratios of simultaneously occurring sounds, we also calculated
the *AkI* parameter:

$$AkI = N_{R1} \cdot N_{R2} \cdot \ldots \cdot N_{Rk}$$ (4.7)

where k is the number of sounds in a given sample.

From the sequence of *AkI* from a musical segment, the array AkI_S can be defined
as:

$$AkI_S = (AkI_1, AkI_2, \ldots, AkI_p)$$ (4.8)

where p is the number of samples collected from a given segment.

4.3.2.4   Process of Sample Collection from a Segment

The moments of sample collection from a musical segment have been defined according to two criteria. The first is the collection of samples at every eighth (Fig. 4.6), and
the second is the collection of samples at every new chord in a segment (Fig. 4.7).

A sequence of *AkD* samples, collected at every eighth in a segment from Etude
Op. 10, No. 5 by F. Chopin is presented in Fig. 4.8 and from Prelude in C minor Op.
28, No. 20 in Fig. 4.9.

These compositions differ in the types of chords and their frequency of change
(Prelude is slow and majestic, while Etude is quite happy and fast). These differences reflect changes in the *AkD* values, seen on the graphs as zigzags for Etude
and a flowing line for Prelude. Figure 4.8 shows a certain recurrence of dissonance
arrangements of chords (chords 4–8 and 20–24).

Fig. 4.6 Process of sample collection from a segment at every eighth

Fig. 4.7 Process of sample collection from a segment at every new chord in a segment

Fig. 4.8 *AkD* for fragment of F. Chopin Etude Op. 10 No. 5

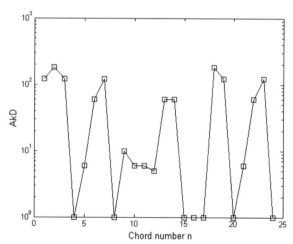

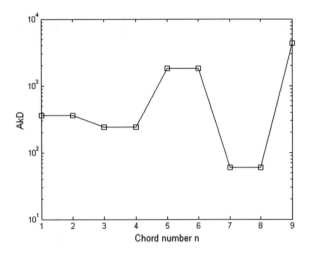

Fig. 4.9 *AkD* for fragment of F. Chopin Prelude in C minor Op. 28, No. 20

4.3.2.5   The List of Harmony Features

Harmony features describe the degree of dissonance in harmony in a given segment (Table 4.3). The statistical features were calculated from sequences AkD_S, AkM_S and AkI_S, which describe the frequency ratios of sounds in chords.

4.3.2.6   Selected Harmony Features and Labeled Emotions

Figure 4.10 presents the spread of values of the Size of AkD_S feature for excerpts labeled by music experts with four emotions. In this case, it is the Size of AkD_S collected at every new chord in a segment. From the box plot we can see that the amount of new chords in music excerpts labeled with emotions e2 and e1 is much higher than for e3 and e4. Also the spread and median value of the amount of new chords in angry (e2) excerpts is higher than for excerpts labeled with other emotions (e1, e3, e4).

Figure 4.11 presents the spread of values of the Second AkD_S Percentage (the percentage share of the second most frequent value in AkD_S) feature for excerpts labeled by music experts with four emotions. The median value of this feature in e3 (sad) and e4 (relaxed) excerpts is a little higher than for e1 (happy) and e2 (angry) excerpts. In this case, we can see that it is not always possible to draw clear conclusions from the spread of values. Nevertheless, we can see that the spreads do not completely overlap and quite possibly in combination with other features the Second AkD_S Percentage could be useful. The usefulness of the Second AkD_S Percentage was confirmed in the feature selection process while building classifiers (Chap. 5 Sect. 5.5.1.2).

Table 4.3 Main harmony features

Feature	Description
Average AkD_S	Average of values in array AkD_S
Standard Deviation of AkD_S	Standard deviation of values in array AkD_S
Median AkD_S	Median value of array AkD_S
Size of AkD_S	Number of samples in array AkD_S
Numerical Integration of AkD_S	Approximate integral of values in array AkD_S
First Max AkD_S	First maximal value in array AkD_S
Second Max AkD_S	Second maximal value in array AkD_S
Third Max AkD_S	Third maximal value in array AkD_S
Average of First 3 Max AkD_S	Average of first 3 max values in array AkD_S
First AkD_S	First most frequent value in array AkD_S
Second AkD_S	Second most frequent value in array AkD_S
Third AkD_S	Third most frequent value in array AkD_S
First AkD_S Percentage	Percentage share of the first most frequent value in AkD_S
Second AkD_S Percentage	Percentage share of the second most frequent value in AkD_S
Third AkD_S Percentage	Percentage share of the third most frequent value in AkD_S
Average AkM_S	Average of values in array AkM_S
Standard Deviation of AkM_S	Standard deviation of values in array AkM_S
Median AkM_S	Median value of array AkM_S
Numerical Integration of AkM_S	Approximate integral of values in array AkM_S
Average AkI_S	Average of values in array AkI_S
Standard Deviation of AkI_S	Standard deviation of values in array AkI_S
Median AkI_S	Median value of array AkI_S
Numerical Integration of AkI_S	Approximate integral of values in array AkI_S

Fig. 4.10 Box plot of the Size of AkD_S feature for MIDI data set labeled with four emotions e1–e4

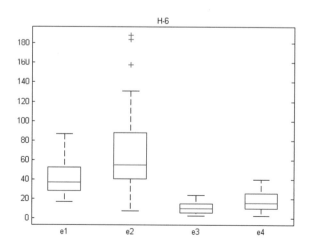

Fig. 4.11 Box plot of the
Second AkD_S Percentage
feature for MIDI data set
labeled with four emotions
e1–e4

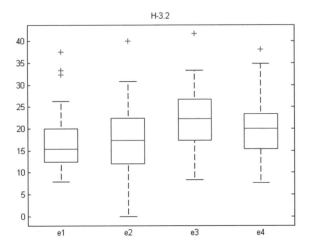

4.3.3   Harmony-Rhythm Features

The moment of appearance of a given accent, chord, dissonance, etc. in the bar is
of great significance. The most important and significant parameters were obtained
taking into account rhythm parameters. We created a special group of harmony features connected with rythm features, which describe the moments in which harmony
features are extracted.

4.3.3.1   Process of Sample Collection from a Segment

We created an AkD_B data table, which comprises of AkD samples collected from
musical segments at moments of the Strongest Pulses (beginnings of bars, repeating
accents that dominate in a given fragment). All values from the beat histogram that
were more than 50% of the First Strongest Rhythmic Pulse in a beat histogram were
accepted as the Strongest Pulses (Sect. 4.3.1).
    AkD_B data table is defined as:

$$AkD_B = (AkD_1, AkD_2, \dots, AkD_b) \tag{4.9}$$

where b is the number of collected samples at moments of the Strongest Pulses.
    The moments of sample collection from a segment using information about the
moments of the Strongest Pulses is presented in Fig. 4.12.
    Statistical features from AkD_B were calculated, just as with AkD_S (Table 4.3). The
same pertained to features obtained from AkM_B and AkI_B.

Fig. 4.12 Process of sample collection from a segment at the moments of the Strongest Pulses

Fig. 4.13 Box plot of the First AkD_B Percentage feature for MIDI data set labeled with four emotions e1–e4

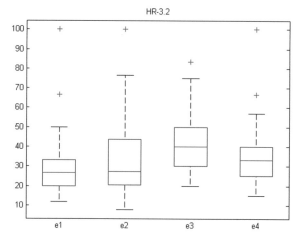

4.3.3.2 Selected Harmony-Rhythm Features and Labeled Emotions

Figures 4.13 and 4.14 present a spread of values of two selected features for excerpts labeled by music experts with four emotions.

From the box plot of the First AkD_B Percentage (Fig. 4.13), we can see that the music excerpts labeled with emotions from the left quarters of Russell's model of emotions (e2, e3) have higher spread values than those from the right quarters (e1, e4). The First AkD_B Percentage feature could be useful for differentiating emotions on the valence axis, between e2 (angry), e3 (sad) and e1 (happy), e4 (relaxed).

When we observe the value distribution of the Numerical Integration of the AkM_B feature for files labeled with four emotions (Fig. 4.14), we notice the higher median values for excerpts labeled with emotion e2 (angry) and the lowest median values for emotion e3 (sad). The Numerical Integration of AkM_B feature could be useful for differentiating emotions on both axes of valence and arousal.

Fig. 4.14 Box plot of the
Numerical Integration of
AkM_B feature for MIDI data
set labeled with four
emotions e1–e4

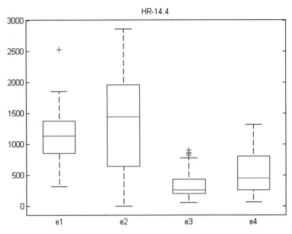

Table 4.4 Main dynamic
features

Feature	Description
Average loudness	Average of loudness levels of all notes
Standard deviation of loudness	Standard deviation of loudness levels of all notes

4.3.4   Dynamic Features

Dynamic features are based on the intensity of sound, the length of the sounds, and
their variability in a segment (Table 4.4). The loudness levels of notes were taken
from MIDI events Note On.

4.4   Conclusions

We presented features created for emotion detection in MIDI files divided into four
groups: rhythm, harmony, harmony-rhythm, and dynamic features. Harmony and
harmony-rhythm are our own solutions. We presented their potential to individually discriminate between emotion categories. Analysis of the value distribution of
selected features for MIDI excerpts labeled with four basic emotions suggests that
some features are more useful for detecting emotions on the valence axis while others are more effective for differentiating emotions on the arousal axis. The ultimate
usefulness of MIDI features will be confirmed in the feature selection process while
building classifiers (Chap. 5 Sect. 5.5.1.2).

   In the presented set, there are no features pertaining to the timbre of sound. In the
studied MIDI set, we only studied piano compositions. We intentionally introduced

this simplification to be able to focus more on which notes are being played and not the tone.

In the future, development and finding new melodic features could be an interesting enhancement of the presented set. Melodic features selecting the main melody and describing its course in time could certainly contribute to even better differentiation of emotions in music excerpts.

Chapter 5
Hierarchical Emotion Detection in MIDI Files

5.1 Introduction

In our research, we concentrated on emotion detection in MIDI files [87] containing symbolic representation of music (key, structure, chords, instrument). The means of representation of music content in MIDI files is much closer to the description that is used by musicians, composers, and musicologists. To describe music, they use key, tempo, scale, sounds, etc. This way, we avoid the difficult stage of extracting separate notes, tracks, and instruments from audio files; and we can concentrate on the deciding element, which is the music content.

Listening to music is a particularly emotional activity. People need a variety of emotions, and music is perfectly suited to provide it to them. However, it turns out that musical compositions do not contain one type of emotion, e.g. only positive or only negative. During the course of one composition, these emotions can take on a variety of shades and change several times with varying intensity.

Apart from emotion detection, this chapter presents a strategy for the analysis of emotions contained within musical compositions. We present a method for tracking changing emotions during the course of a musical piece. The collected data allowed determining the dominant emotion in the musical compositions, presenting emotion histograms, and constructing maps visualizing the distribution of emotions over time.

5.2 Related Work

There are few papers dedicated to emotion detection in MIDI files; most focus on emotion detection in audio files [49, 118]. In addition to studies on emotion detection, there are papers on modifying MIDI file parameters with the aim of obtaining a specified emotion.

© Springer International Publishing AG 2018
J. Grekow, *From Content-Based Music Emotion Recognition to Emotion Maps of Musical Pieces*, Studies in Computational Intelligence 747,
https://doi.org/10.1007/978-3-319-70609-2_5

Wang et al. in [71] applied a hierarchical model for emotion detection. Emotion groups were created on the basis of Thayer's model and contained 2 emotions at the first level and 6 emotions at the second. The features used to build classifiers referred to pitch, intervals, tempo, instrument type, meter, and tonality.

Emotion detection in MIDI files can also be found in the work of DiPaola and Arya [17], who combined the emotional content of a piece with visualization elements. They used the detected emotion for animating a 3-D face. The features used referred to rhythm, volume, timbre, articulation, melody, and tonality.

Lin et al. [60] examined music emotion regression performance using audio, lyric, and MIDI features. Two sets of MIDI files were used: the first set was converted from audio files, and the second set was obtained from the Internet and musical score conversion. They found that the MIDI features performed better than the audio features.

A connection between MIDI files and emotion was presented in [9], where a computer program was used to produce performances with different emotional expressions. The program used a set of rules characteristic for each emotion (fear, anger, happiness, sadness, solemnity, tenderness), which were used to modify such parameters of MIDI files as tempo, sound level, articulation, tone onsets and delays.

Livingstone and Brown [61] proposed a dynamic music environment, where MIDI music tracks adjusted in real-time to the emotion in the computer games. Music emotion rules, which connect 8 emotion categories to musical elements such as mode, tempo, loudness, harmonic complexity and articulation, were collected and implemented.

Moriguchi et al. [69] proposed a system for controlling the degrees of emotions in MIDI files. Parameters such as timbre, tempo, number of performance tracks, and loudness of a given excerpt were used to modify the expressed emotion in the music when played back to the listener.

5.3   MIDI Music Data

The data set that was used in the conducted experiments consisted of 350 six-second MIDI excerpts and was described in detail in Chap. 3 Sect. 33.4. The hierarchical emotion model we used (Fig. 5.1) was based on Russell's circumplex model, and consisted of emotion categories on two levels, L1 and L2. The first level (L1) contains 4 categories, while the second level (L2) is related to the first, and is made up of 12 sub-emotions, 3 emotions for each emotion contained in the first level.

Data annotation was done by five music experts with a university music education. The amount of obtained examples labeled by emotions on the first level are presented in Table 5.1, and those labeled by emotions on the second level are presented in Table 5.2.

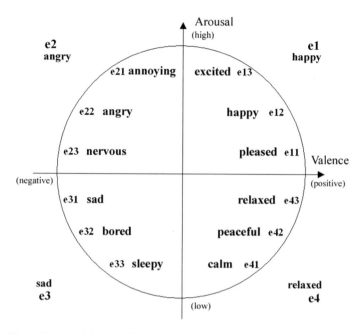

Fig. 5.1 Hierarchical model of emotions based on Russell's circumplex model

Table 5.1 Amount of MIDI examples labeled by emotions from the first level (L1)

Emotion abbreviation	Emotion	Amount of examples
e1	happy	89
e2	angry	105
e3	sad	79
e4	relaxed	77

5.4 Feature Extraction

We used our own software (written by the author) for feature extraction. 63 MIDI features were obtained for each 6-second labeled MIDI excerpt, and the extracted MIDI features were divided into four groups: rhythm, harmony, harmony-rhythm, and dynamic, and were described in Chap. 4 Sect. 4.3.

Table 5.2 Amount of MIDI examples labeled by emotions from the second level (L2)

Emotion abbreviation	Emotion	Amount of examples
e11	pleased	39
e12	happy	36
e13	excited	14
e21	annoying	35
e22	angry	36
e23	nervous	34
e31	sad	37
e32	bored	23
e33	sleepy	19
e41	calm	19
e42	peaceful	15
e43	relaxed	43

5.5 Construction of Classifiers

5.5.1 First Level Classifiers

5.5.1.1 One Classifier Recognizing Four Emotions

The classifier on the first level should be able to recognize the 4 categories of emotions that correspond to the four quarters of Russell's model: happy, angry, sad, and relaxed. From the music samples labeled by music experts and the extracted MIDI features, we created ARFF files that were the input data for the algorithms building classifiers.

We built classifiers for emotion detection using the following algorithms: J48, BayesNet, K-nn, SMO. J48 implements the C4.5 decision tree [82], BayesNet is an implementation of the Bayesian network classifier [41], K-nn represents K-nearest neighbours classifier [1], and SMO implements sequential minimal optimization algorithm for training a support vector classifier [78].

The classification results were calculated using a cross validation evaluation CV-10. The highest accuracy among all tested algorithms was obtained for the J48 algorithm (Table 5.3).

Table 5.3 Accuracy of classifiers obtained for first level (L1) classifiers

Classifier(%)	J48	BayesNet	K-nn	SMO
Accuracy after attribute selection	**76.00**	70.28	74.28	66.85
Accuracy after attribute selection	**82.00**	70.00	81.42	73.42

Table 5.4 Confusion matrix obtained for the J48 algorithm

		Predicted class			
		e1	e2	e3	e4
Actual class	e1	**66**	17	1	5
	e2	18	**83**	3	1
	e3	0	0	**71**	8
	e4	2	2	6	**67**

Classifier accuracy for algorithm J48 improved to **82.00 %** after applying attribute selection (attribute evaluator: Wrapper Subset Evaluator [50], search method: Best First [117]).

From the confusion matrix (Table 5.4) obtained during classifier evaluation, we can conclude that usually fewer mistakes are made between the top (e1, e2) and bottom (e3, e4) quadrants of Russell's model. At the same time, errors in differentiating emotions on the valence axis, between emotions e1 and e2, and between e3 and e4 are significantly more frequent.

The most important features for the detection of emotions on the first level (L1), selected from a rich set of features, presented in Chap. 4 Sect. 4.3, were:

- Size of AkD_S at every eighth (H),
- Median AkD_S at every eighth (H),
- Size of AkD_S at every new chord in a segment (H),
- Second Max AkD_B at every eighth (HR),
- Numerical Integration of AkM_B (HR),
- Median AkI_B (HR),
- First Strongest Rhythmic Pulse—SRP1 (R),

where H, HR, and R represent feature group abbreviations: harmony (H), harmony-rhythm (HR), and rhythm (R).

The selected features are mainly harmony and harmony-rhythm features. They present the statistics collected from a sequence of values as Chord Degree of Dissonance (AkD). Harmony-rhythm features pertain to statistics collected from sequences of parameters AkD, AkM, AkI, which describe the frequency ratios in chords, collected from musical segments at moments of the Strongest Pulses. They confirm the usefulness of the created features. In the selected features, we have representatives of all groups except for dynamic features, which in most likelihood were covered by other features. The First Strongest Rhythmic Pulse (the beat with the highest magnitude in the beat histogram) is also an important feature, which is a logical explanation of the effect of rhythm on the detected emotion.

Table 5.5 Accuracy obtained for High-Low Arousal and High-Low Valence classifiers using J48

Classifier	High-Low	High-Low
Accuracy (%)	**97.42**	82.57

5.5.1.2   Two Classifiers Recognizing Emotions on the Arousal and Valence Axes

To find which of the MIDI features are better suited for emotion differentiation on the arousal and valence axes from Russell's emotion model, we build two additional classifiers. The first one (High-Low Arousal) had the task of differentiating emotions from the top part of the semicircle of the model (e1, e2) from emotions from the bottom part of the semicircle (e3, e4). The second one (High-Low Valence) had to differentiate emotions from the left part of the semicircle (e2, e3) from emotions on the right part of the semicircle (e1, e4).

The classifiers were built using the J48 algorithm, which was the winner during building one classifier for the four emotions of L1. The classification results were calculated using a cross validation evaluation CV-10. For attribute selection we used attribute Wrapper Subset Evaluator and search method Best First. The accuracy obtained for the J48 algorithm after using attribute selection is presented in Table 5.5.

The high accuracy for the High-Low Arousal classifier (97.42%) indicates that the classifier, using the collected MIDI features, differentiates emotions in the top half from the bottom half of the model well. A slightly worse result (82.57%) was obtained for the High-Low Valence classifier, which confirms that recognizing emotions on the valence axis is more difficult than on the arousal axis.

Table 5.6 presents the most important features for detecting emotions on the arousal and valence axes. Both sets have selected rhythm features (R) describing the number of strong pulses (Relatively Strong Pulses) obtained from the beat histogram. We can also see the usefulness of features pertaining to the duration of notes (Note Duration) or number of notes per second (Note Density). It is interesting that in the feature set for the detection of High-Low Valence we only have harmony-rhythm features (HR) and not harmony features (H). We found that statistics from harmony features collected from musical segments at moments of the Strongest Pulses are more useful for detecting High-Low Valence than harmony features collected at every eighth (H). During High-Low Arousal detection, most harmony features (H) collected at every eighth note were enough.

Table 5.6 Selected features used for building High-Low Arousal and High-Low Valence classifiers

Classifier	Selected features
High-Low Arousal	Size of AkD_S at every eighth (H)
	First AkD_S at every eighth (H)
	Second AkD_S Percentage at every eighth (H)
	Third AkD_S at every eighth (H)
	Second AkD_B Percentage (HR)
	Relatively Strong Pulses 10 (RSP10) (R)
	Average Note Duration (R)
	Standard Deviation of Note Duration (R)
High-Low Valence	Numerical Integration of AkD_B (HR)
	First AkD_B Percentage (HR)
	Average of First 3 Max AkD_B (HR)
	Median AkM_B (HR)
	Average AkI_B (HR)
	Relatively Strong Pulses 30 (RSP30) (R)
	Note Density (R)

5.5.2   Second Level Classifiers

5.5.2.1   One Classifier Recognizing Twelve Emotions

While building the second level classifiers, we first decided to build a classifier that would differentiate 12 sub-emotions. We built classifiers for emotion detection using the following algorithms: J48, BayesNet, K-nn, SMO. The classification results were calculated using a cross validation evaluation CV-10.

Once again, we obtained the best accuracy for the studied algorithms with J48, with a value of 65.14% (Table 5.7); although the accuracy was lower by 17 % points than the accuracy of the classifier detecting 4 emotions on the first level (L1). In the case of detecting 12 sub-emotions, the classifier is clearly less accurate, which is connected with the greater number of classes with the same number of features. Also, the quality of data labeling by the experts at level L2 is generally lower than for level L1, which may also lower the results.

Table 5.7 Accuracy of classifiers obtained for second level (L2) classifiers

Classifier(%)	J48	BayesNet	K-nn	SMO
Accuracy before attribute selection	**60.00**	56.00	54.28	52.28
Accuracy after attribute selection	**65.14**	58.00	63.55	60.00

Table 5.8 Four classifiers for the second level (L2)

Name of classifier	Detected emotions of second level
CL21	e11, e12, e13
CL22	e21, e22, e23
CL23	e31, e32, e33
CL24	e41, e42, e43

Table 5.9 Accuracy of classifiers obtained for 4 second level (L2) classifiers

Name of classifier(%)	CL21	CL22	CL23	CL24
Before attribute selection	76.40	74.42	70.88	80.51
After attribute selection	84.76	84.76	84.81	**93.50**

5.5.2.2   Four Classifiers Recognizing Sub-emotions

In order to improve emotion detection accuracy on the second level (L2), we decided
to build 4 classifiers, one for each quarter of Russell's emotion model (Table 5.8).
Each of the classifiers specializes in detecting 3 sub-emotions for the respective
quarter, CL21—detects emotions in the first quarter, CL22 in the second, CL23 in
the third, and CL24 in the fourth.

To create the specific classifiers for level L2, we used samples from a given
category. In other words, to build classifier CL21 for emotions e11, e12, and e12
we only used sampled labeled as e1 at level L1. We did the same for the remaining
classifiers, CL22, CL23, and CL24.

Thus, we obtained 4 classifiers specializing in detecting sub-emotions at level L2.
To build the classifiers, we used algorithm J48, which was the winner when we built
classifiers at levels L1 and L2. The obtained accuracy for each classifier before and
after attribute selection is presented in Table 5.9.

Notice the clear improvement in accuracy (84.74–93.50%) compared with the
accuracy obtained for the classifier detecting 12 emotions at level L2 (65.14%). These
results confirm the usefulness of the classifiers and that 4 classifiers specializing in
detecting 3 sub-emotions for each quarter of the model detect emotions better than
one classifier detecting all 12 emotions.

5.6   Hierarchical Classification

Level L1 and L2 classifiers were used for hierarchical emotion detection in music
files (Fig. 5.2). Emotion detection in music files was done analogous to the used
hierarchical model of emotions with categories on two levels L1 and L2.

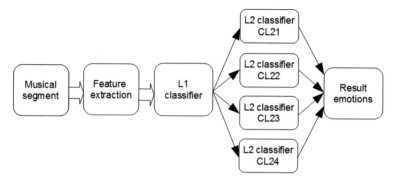

Fig. 5.2 Hierarchical emotion detection in a musical segment

First, a musical segment underwent feature extraction; then, the obtained features vector representing the musical segment was classified at level L1. One classifier (Sect. 5.5.1.1) detecting 4 emotions—e1—happy, e2—angry, e3—sad, e3—relaxed—was used. Next, depending on the results of the first classifier, the appropriate level L2 classifier was selected (Sect. 5.5.2.2); its task was to detect 3 sub-emotions. For example, if at the first level the detected emotion was e1, then at the second level a classifier detecting sub-emotions for e1, i.e. classifier CL21, was used. If the result of classification at level L1 was emotion e2, then at level L2 we used classifier CL22. And so on for the remaining cases. The result of hierarchical classification of musical segments was the detection of emotions on two levels, L1 and L2.

5.7 Emotion Tracking in MIDI Files

5.7.1 System Construction

The proposed system for tracking emotions in a musical composition is shown in Fig. 5.3. It consists of a database of musical compositions, composition segmentation, hierarchical emotion detection, and the result presentation module. The resulting emotion labels were used to designate the consecutive segments of a musical composition. The collected data allowed for the analysis of a musical composition in terms of the emotions contained therein.

When using the system, the user first selects a musical composition from the database, then cross-indexes it for emotion. Finally, an analysis and visualization of the obtained results are conducted.

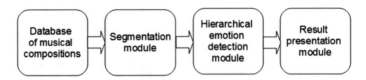

Fig. 5.3 Construction of the emotion tracking system

5.7.2 Musical Composition Segmentation

Emotions in musical compositions are not constant. In a fragment lasting several seconds, there may be just one emotion or it may change many times. The emotion can change in very different ways over the course of a musical composition lasting only a few minutes depending on the musical content of the piece. Emotion reflects what is happening in the musical composition, for example, if the pace of the composition increases, the emotion changes in the direction of the upper quadrants (e1, e2) of Russell's model. If the sounds of the piece begin to be less consonant, the expressed emotions come from the left lateral quadrants (e2, e3) of Russell's model.

Some pieces may have many emotional changes (e.g. musical compositions of varying moods or affecting the listener with a whole range of musical means such as different pace, variable rhythm, dynamics, etc.), while others may be based on one unchanging emotion (e.g. musical compositions of uniform structure with a steady pace, dynamics, and rhythm).

Detection of emotion was conducted in our research on 6-second segments, with each consecutive segment shifted by 2 s; thus, successive segments overlapped at a 2/3 ratio (Fig. 5.4). This allowed exactly tracking and detecting even the slightest

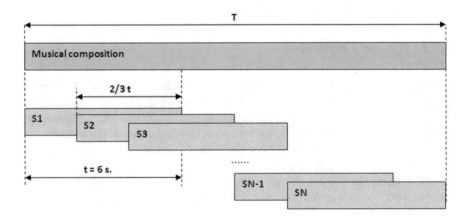

Fig. 5.4 Division of a musical piece into segments

change of emotion in the examined musical composition. For a musical composition
lasting $T = 120$ s, $N = 60$ segments ($S1$, $S2$, ..., $S59$, $S60$) were analyzed, and for
each L1 and L2 level of emotion detection was performed.

5.8   Results of Emotion Tracking in MIDI Files

The result of tracking emotions of a musical composition are segments with emotions
described on two levels: the higher, more general, L1; and the lower, more detailed,
L2. Analysis of the obtained emotions confirms the assumption that emotions are
not uniform in a musical piece.

5.8.1   Emotion Histograms of Musical Compositions

Emotions can change throughout a musical composition. Some emotions are more
common than others and their type is not always the same. The first method used
for presenting the distribution of emotions in a musical composition is emotion
histograms. Figures 5.5 and 5.6 present the emotion histograms of two compositions:
Ludwig van Beethoven's Sonata No. 23 F minor, Opus 57, part 1 (Appassionata),
and Frédéric Chopin's Prelude in C minor Op.28, No. 20. On the presented graphs,
the horizontal axis corresponds to the type of emotion, and the height of the bar
indicates how often a specific emotion occurred.

  Figure 5.5a presents the histogram of L1 level emotions in Beethoven's Appassionata sonata, in which emotion e2 (angry) occurs in 76% of the segments and is
dominant. The second, most significant, emotion is e1 (happy), which occurs in 20%

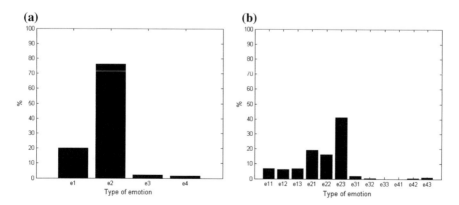

Fig. 5.5 Histogram of L1 (**a**) and L2 (**b**) level emotions in L.v. Beethoven's Sonata No. 23 F minor,
Opus 57 (Appassionata), part 1

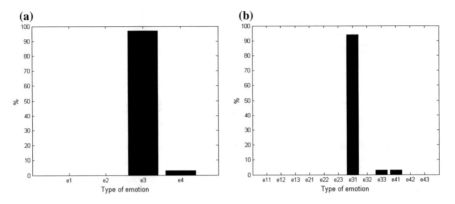

Fig. 5.6 Histogram of L1 (**a**) and L2 (**b**) level emotions in F. Chopin's Prelude in C minor Op.28, No. 20

of the segments. We notice that the occurrence of emotions e4 (relaxed) and e3 (sad) is very rare in the given composition and occur in about 2% of the segments each.

Figure 5.5b presents the histogram of L2 level emotions in Beethoven's Appassionata sonata. Analyzing it and comparing it with the L1 level histogram (Fig. 5.5a), we can see in detail how an emotion from level L1 (e2) breaks down into emotions from level L2: e21 (annoying), e22 (angry), and e23 (nervous). Notice the domination of emotion e23 (40%). The sub-emotions of the first quarter of Russell's model (e1) are e11 (pleased), e12 (happy), and e13 (excited), and they occur in about 7% of segments each.

The contrast to the presented histograms of Beethoven's Appassionata (Fig. 5.5) is the histograms of F. Chopin's Prelude in C minor Op.28, No. 20 (Fig. 5.6). Not only is there a different main emotion, its domination is much greater. We can notice a great domination of the main emotion from level L1, e3 (97%), and the domination of emotion e31 (94%) from level L2. The occurrence of other emotions is marginal.

Emotional diversity is much richer in Appassionata; there are two main emotions, e1 and e2, from level L1, with various shades of emotions at level L2. In Prelude, we have one dominating emotion, e3, from level L1 and one, e31, from level L2. In other words, there is a lack of diversity in shades of emotions.

5.8.2 Emotion Maps

Another method used to analyze emotions in a musical composition is detailed maps showing the distribution of emotions for the duration of a piece (Figs. 5.7, 5.8, 5.9 and 5.10). The horizontal axis shows the time in seconds and the vertical axis the emotions occurring at a given moment.

In Fig. 5.7, presenting a map of L1 level emotions for Beethoven's Appassionata, we notice that e2 is dominant throughout the entire piece. From the map, we can see

Fig. 5.7 Map of L1 level
emotions in L.v. Beethoven's
Sonata No. 23 F minor, Opus
57 (Appassionata), part 1

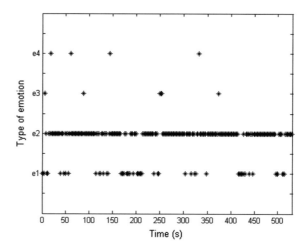

Fig. 5.8 Map of L2 level
emotions in L.v. Beethoven's
Sonata No. 23 F minor, Opus
57 (Appassionata), part 1

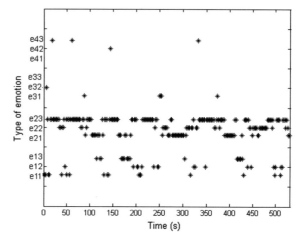

when the second most frequent emotion, e1, occurs. The occurrence of emotions e3
and e4 is incidental.

By analyzing the map of L2 level emotions for L.v.Beethoven's Appassionata
(Fig. 5.8), we can notice the detailed distribution of emotions. The set of occurring
emotions is quite rich. One could make an attempt to find patterns in the presented
map, for example, we noticed the subsequent occurrence of emotions e23, e22, e21
forming falling 'stairs' in several moments of the piece: s. 70–100, s. 240–260,
s. 360–390.

The contrast to the presented maps of Beethoven's Appassionata is the maps of
F. Chopin's Prelude in C minor Op.28, No. 20 (Figs. 5.9 and 5.10). The dominating
emotion, e3, throughout the entire composition from level L1 is presented in the
form of a horizontal line. A short change in emotions to e4 occurs only in the 44th
second (Fig. 5.9). A similar horizontal line for e31 occurs at level L2 (Fig. 5.10).

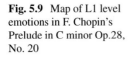

Fig. 5.9 Map of L1 level emotions in F. Chopin's Prelude in C minor Op.28, No. 20

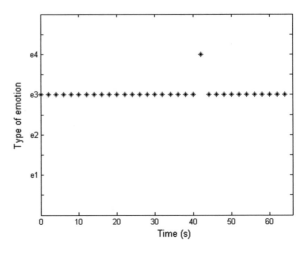

Fig. 5.10 Map of L2 level emotions in F. Chopin's Prelude in C minor Op.28, No. 20

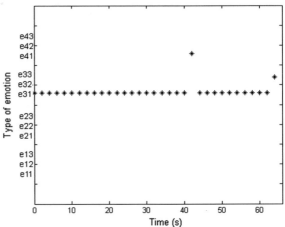

Comparing the maps of the two compositions, we can see how much they vary in their distribution of emotions over time.

5.8.3 Quantity of Changes of Emotion

Because some pieces may have many emotional changes (e.g., songs with varying moods), while others may be based on a single dominant emotion (e.g. musical compositions with a steady pace, dynamics and rhythm, etc.), we introduced the Quantity of Changes of Emotion (QCE) in a musical composition, which is the sum of the number of changes of emotion in adjacent segments. To make the indicator

Table 5.10 The dominant emotion and the Quantity of Changes of Emotion (QCE) in a piece

Piece	QCE L1	QCE L2	Dominating emotion in L1 (percentage)	Dominating emotion in L2 (percentage)
Appassionata, part 1	24.34	39.33	e2 (76%)	e23 (40%)
Prelude Op.28, No. 20	5.88	8.82	e3 (97%)	e31 (94%)

values independent from the length of the piece, the obtained sum was divided by the number of N segments.

$$QCE = \frac{\sum_{i=1}^{N-1} f(i)}{N} * 100 \tag{5.1}$$

$$f(i) = \begin{cases} 1, & \text{if } Emotion(i) \neq Emotion(i+1) \\ 0, & \text{if } Emotion(i) = Emotion(i+1) \end{cases} \tag{5.2}$$

where i is the number of the segment in the piece, N the number of segments in the composition, and $Emotion(i)$ represents the emotion of the i segment. The function $f(i)$ indicates whether the adjacent segments have a different (value 1) or same (value 0) emotion. The more changes of emotion in a musical composition, the greater the QCE value.

Table 5.10 presents the obtained values for the quantity of changes of emotion and the dominating emotions for the presented two compositions. We can notice that the dominant emotion percentages are much higher in the Prelude than in the Appassionata. Also, the QCE in the Prelude has smaller values than the Appassionata at both levels L1 and L2. From the obtained results, we can conclude that the Prelude is more emotionally homogeneous with a greater dominance of individual emotions and Beethoven's Appassionata is more diverse emotionally.

The method of creating emotions in a musical piece depends on the composer. These emotions can be presented in the forms of histograms, maps, or using such parameters as QCE. A search for parameters describing the emotional distributions in compositions could be an interesting continuation of this work in the future.

5.9 Conclusions

In this chapter we presented emotion detection in pieces of classical music in the form of MIDI files. A hierarchical model of emotions consisting of two levels, L1 and L2, was used. A collection of harmony and rhythm MIDI features extracted from music files allowed for emotion detection with an average of 82% accuracy at level L1. The built classifiers detecting emotions at level L2, i.e. the sub-emotions of level L1, achieved accuracy between 84 and 93%. They were built so that they specialize in detecting emotions from a selected group of sub-emotions, which improved their effectiveness.

During feature selection, we found the most useful MIDI features to build a classifier recognizing four emotions on the first level. We also found the most important features to distinguish emotions on the arousal and valence axes from Russell's emotion model; harmony-rhythm MIDI features proved to be particularly useful for emotion detection on the valence axis.

A strategy for the analysis of emotions contained within MIDI musical compositions was presented. We constructed the system for tracking changing emotions during the course of a musical piece, and the collected data allowed determining the dominant emotion in the musical compositions, presenting emotion histograms, and constructing maps visualizing the distribution of emotions in time.

The amount of changes of emotions during a piece may be different; therefore, we introduced a parameter evaluating the quantity of changes of emotions in a musical composition. The information obtained about an emotion in a piece made it possible to analyze the musical compositions, thus providing new knowledge about the compositions and the method of their emotional development.

Part III
Emotion Detection in Audio Files

Chapter 6
Audio Features

6.1 Introduction

In order to detect emotions in music files, we first need to perform extraction of audio features that describe/represent the music content comprising the emotional content of music. Audio features are extracted from audio signals, which can be presented in the time domain as well as the spectral domain. Extracted features should refer to those elements of music that affect the creation of emotions: timbre, dynamic, rhythm, harmony, melody, etc. Describing the rich music content using audio features is not a simple task, and the currently available feature set is quite rich, although it still cannot completely describe the complex structure of music on the one hand, nor the influence of these structural elements on perceived emotions on the other. Very rich feature sets only come close to complete description of music content; they are a partial approximation.

6.2 Features from Audio Analysis Tools

Current audio analysis tools tailored for Music Information Retrieval (MIR), such as Essentia [8], Marsyas [106], jAudio [65], PsySound [11] and MIRtoolbox [53], contain a number of executable extractors computing music descriptors for an audio track: spectral, time-domain, rhythmic, and tonal descriptors. Below, we present two tools, Essentia and Marsyas, which were used for feature extraction in the conducted experiments described herein.

© Springer International Publishing AG 2018
J. Grekow, *From Content-Based Music Emotion Recognition to Emotion Maps of Musical Pieces*, Studies in Computational Intelligence 747,
https://doi.org/10.1007/978-3-319-70609-2_6

6.2.1 Essentia

Essentia [8] is an open-source C++ library, which was created at Music Technology Group, Universitat Pompeu Fabra, Barcelona. Essentia (version 2.1beta) contains a number of executable extractors computing music descriptors for an audio track: spectral, time-domain, rhythmic, tonal descriptors; and returning the results in YAML and JSON data formats. Extracted features by Essentia are divided into three groups: low-level, rhythm, and tonal features (Table 6.1).

Essentia also calculates many statistic features: the mean, geometric mean, power mean, median of an array, and all its moments up to the 5th-order, its energy, and the root mean square (RMS). To characterize the spectrum, flatness, crest and decrease of an array are calculated. Variance, skewness, kurtosis of probability distribution, and a single Gaussian estimate were calculated for the given list of arrays.

Table 6.1 The feature set obtained from Essentia audio analysis tool

Group	Features
Low-level	Loudness
Timbre features	Barkbands
	Erbbands
	Melbands
	Mel Frequency Cepstral Coefficients (MFCC)
	Spectral: Complexity, centroid, decrease, crest, flatness, kurtosis, skewness, spread, flux, rolloff
	High frequency content
	Pitch salience
	Silence rate
	Dissonance
	Zero crossing rate
Mid-level	BPM histogram
Rhythm features	Beats loudness
	Beats loudness band
	Danceability
	Onset rate
Mid-level	Chords histogram
Tonal features	Chords changes rate
	Chords number rate
	Chords strength
	Harmonic pitch class profile,
	Key strength
	Mode

Feature sets obtained from Essentia, along with their calculated statistics, can create large vectors containing more than 500 features. A full list of features is available on the web.[1]

6.2.2 Marsyas

Marsyas (Music Analysis, Retrieval and Synthesis for Audio Signals) is an open source software framework for audio processing software, written by George Tzanetakis [106]. It is implemented in C++ and retains the ability to output feature extraction data to ARFF format. With this tool, the following features can be extracted:

- Zero Crossings Rate,
- Spectral Centroid,
- Spectral Flux,
- Spectral Rolloff,
- Mel Frequency Cepstral Coefficients (MFCC)—13 features,
- Chroma—14 features,

- 31 features in total.

For each of these basic features, Marsyas calculates four statistic features:

1. *The mean of the mean* (calculate mean over the 20 frames, and then calculate the mean of this statistic over the entire segment);
2. *The mean of the standard deviation* (calculate the standard deviation of the feature over 20 frames, and then calculate the mean of these standard deviations over the entire segment);
3. *The standard deviation of the mean* (calculate the mean of the feature over 20 frames, and then calculate the standard deviation of these values over the entire segment);
4. *The standard deviation of the standard deviation* (calculate the standard deviation of the feature over 20 frames, and then calculate the standard deviation of these values over the entire segment).

In this way, we obtained 124 features.

6.3 Selected Audio Features and Labeled Emotions

In this paragraph, we describe selected, relevant features for this emotion classification. Visualizations of distributions of these features between four basic emotions (e1—happy, e2—angry, e3—sad, e4—relaxed) are based on the data set labeled by music experts (Chap. 3 Sect. 3.3).

---

[1]http://essentia.upf.edu/documentation/algorithms_reference.html.

6.3.1 Timbre Features

6.3.1.1   Mel Frequency Cepstral Coefficients

Mel Frequency Cepstral Coefficients (MFCCs) are often used in audio analysis, particularly speech research and music classification tasks. The process of creating MFCC features can be separated into several steps [62, 84]. The first step is to divide signal frames by applying a windowing function (typically Hamming window). In the next step, we compute the Fast Fourier Transform for each frame. Because perceived loudness of a signal is approximately logarithmic, we take the logarithm of the amplitude spectrum. The phase information is discarded. In the next step, we convert the spectrum into the perceptual-based Mel spectrum and divide the spectrum into bands. The linear frequency axis is converted into the Mel scale, and the Mel scale is divided into equally spaced bands. On the Mel scale, lower frequencies are more important than higher frequencies. The relation between the frequency domain and the Mel scale is presented in the following equation:

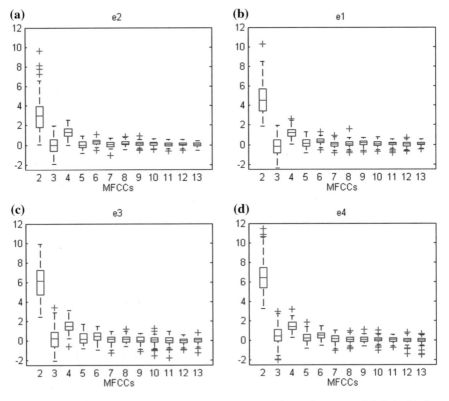

Fig. 6.1 Box plots of the mean values of the MFCC 2–13 in music excerpts labeled with four emotions e1–e4

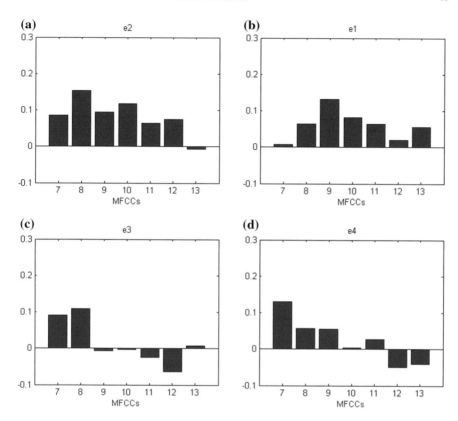

Fig. 6.2 The mean values of the MFCC 7–13 in music excerpts labeled with four emotions e1–e4

$$M = 2595 \cdot \log_{10}(1 + \frac{f}{700}) \qquad (6.1)$$

where f is the frequency in Hz, and M is the frequency in the Mel scale.

The energy of the bins of the FFT of each Mel band is summed. Because components that represent the energy of each Mel band are highly correlated and to reduce the number of parameters that describe a frame, a discrete cosine transformation is applied. Using this transform 13 cepstral coefficients are obtained for each frame. Coefficient number 1 is not taken into account, because it is proportional to the energy.

When we observe the value distribution of MFCCs for files labeled with four emotions (Fig. 6.1), we notice differences in the median values of MFCC No. 2. For excerpts labeled with emotions e3, e4 (low arousal, Fig. 6.1c, d), this coefficient has higher values (over 6) than those labeled with emotions e1, e2 (high arousal, Fig. 6.1a, b). The lowest median value is noticed for excerpts labeled with e2.

When we look closer at the mean values of coefficients with higher numbers, MFCCs No. 7-13 (Fig. 6.2), we see that for each group of excerpts labeled with four emotions, we have different values for each coefficient. We can notice that coefficients for emotions e3, e4 (low arousal, Fig. 6.2c, d) have more negative values than for emotions e1, e2 (high arousal, Fig. 6.2a, b). Noticing general differences on the valence axis (between emotions e1, e4 and e2, e3) is harder, although they are noticeable.

From observation of mean values of MFCCs, we notice their usefulness in differentiating emotions on the arousal and valence axes, although they would most likely be more effective at differentiating emotions on the arousal axis.

6.3.1.2   Bark Bands Energy

Bark bands are proposed by Zwicker in [121]. They model an approximation of the human auditory system and correspond to the first 24 critical bands of hearing. Below 500 Hz the Bark scale is more linear, and above 500 Hz this scale is near to a logarithmic frequency axis. The relation between the frequency domain and the Bark scale is presented in the following equation:

$$B = 13 \cdot \arctan\left(0.00076 \cdot f\right) + 3.5 \cdot \arctan\left(\frac{f}{7500}\right)^2 \qquad (6.2)$$

where f is the frequency in Hz, and B is the frequency in Bark scale.

By counting the energy of the Bark bands, the linear frequency axis is converted into the Bark scale, which is divided into equally spaced 24 bands. To obtain energy in the Bark bands, we compute the Fast Fourier Transform (FFT) for each frame. This transformation changes the domain from time to frequency. Finally, the energy of the bins of the FFT of each Bark band is summed.

If we compare the mean energy in Bark bands in excerpts labeled with four emotions (Fig. 6.3), we will notice that the excerpts labeled with emotions with high arousal: e1, e2 (Fig. 6.3a, b) differ greatly from excerpts labeled with emotions with low arousal: e3, e4 (Fig. 6.3c, d). A change in valence does not greatly affect distribution; distributions of e1 and e2 are similar, and distributions of e3 and e4 are also similar. Thus, it seems that the energy in Bark bands is good at differentiating emotions on the arousal axis and not that useful on the valence axis.

6.3.1.3   Spectral Centroid

Spectral Centroid is the "center of mass" of the spectrum and represents the brightness of a sound. This feature in defined in Eq. 6.3.

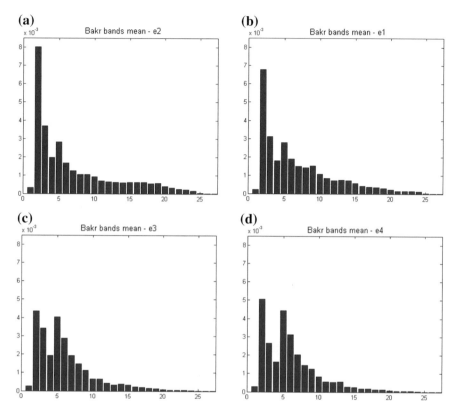

Fig. 6.3 Bar graph of the mean energy in Bark bands in music excerpts labeled with four emotions e1–e4

$$SC = \frac{\sum\limits_{n=0}^{N-1} f(n) \cdot a(n)}{\sum\limits_{n=0}^{N-1} a(n)} \qquad (6.3)$$

where $f(n)$ represents the center frequency of FTT bin number n, and $a(n)$ the amplitude of that bin.

In Fig. 6.4, the spread (space between the first and third quartiles) for excerpts labeled as e2 is higher than for excerpts labeled with other emotions. The median values of spectral centroid enable sorting the four emotions in a descending order, as follows: e2, e1, e3, e4.

Fig. 6.4 Box plot of the spectral centroid mean values for data set labeled with four emotions e1–e4

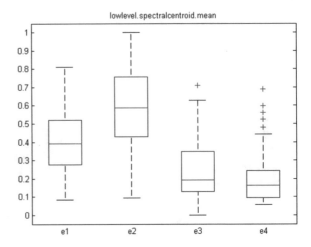

6.3.1.4   Dissonance

Dissonance, also called Harmonic Spectral Deviation in MPEG-7 standard [74, 91], is the deviation of the amplitude harmonic peaks from the global spectral envelope. This feature is computed using the formula:

$$HDEV = \frac{1}{H} \cdot \sum_{h=0}^{H-1} (a(h) - SE(h)) \tag{6.4}$$

where H is the total number of harmonics, $a(h)$ is the amplitude of harmonic h, and $SE(h)$ the amplitude of spectral envelope evaluated at the frequency.

Fig. 6.5 Box plot of normalized dissonance mean values for data set labeled with four emotions e1–e4

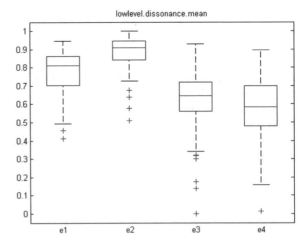

Consonant sounds have more regularly spaced harmonic peaks, the value of $HDEV$ will be lower; and dissonant sounds have more irregularly spaced harmonic peaks, the value of $HDEV$ will be higher.

If we compare $HDEV$ distribution for four emotions (Fig. 6.5), we see that musical examples labeled with e2 (angry) are much more dissonant than other emotions. Median dissonance is much lower than in examples labeled with other emotions. The lowest median dissonance values are observed in examples labeled with emotion e4 (relaxed).

6.3.1.5 Zero Crossing Rate

The zero crossing rate (ZCR) represents the rate at which the number sign changes along a signal. This feature provides an indication of signal noisiness, and is computed for each frame as follows:

$$ZCR = \frac{1}{2} \cdot \sum_{n=1}^{N} |sign(x[n]) - sign(x[n-1])| \qquad (6.5)$$

where the $sign$ function is equal to 1 for positive arguments and 0 for negative arguments, $x[n]$ is the input time domain data, and N is the number of samples in the frame. Because ZCR increases with increased signal noise levels, when using ZCR for emotion detection it would be ideal if the music files were small or had equal noisiness in each category.

It is interesting that this simple indicator is quite good at differentiating excerpts labeled with four basic emotions (Fig. 6.6). Musical examples labeled with emotions from the higher quarters of Russell's model (e1, e2) have on average higher median ZCR than emotions from the lower quarters (e3, e4). Also, the interquartile range

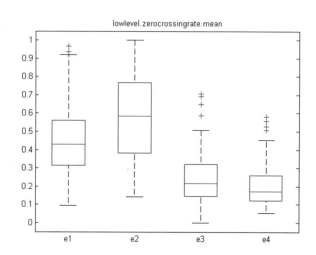

Fig. 6.6 Box plot of normalized zero crossing rate mean value for data set labeled with four emotions e1–e4

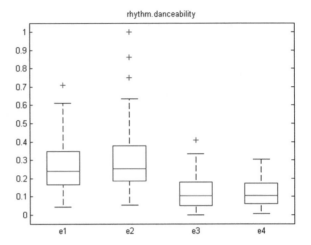

Fig. 6.7 Box plot of the normalized danceability feature for data set labeled with four emotions e1–e4

(IQR = Q3 − Q1), where we have 50% of observations, contains much higher values for emotions e1 and e2 than e3 and e4.

6.3.2   Rhythm Features

6.3.2.1   Danceability

Feature danceability is calculated for the input audio signal; and higher values of this feature mean that the music is more danceable. The algorithm [101] uses Detrended Fluctuation Analysis (DFA) proposed by Peng et al. [75].

From the box plot of danceability for four emotions (Fig. 6.7), we can see that the music excerpts labeled with emotions from the higher quarters of Russell's model of emotions (e1, e2) are more danceable than those from the lower quarters (e3, e4). This means that the danceability feature could be useful for differentiating emotions on the arousal axis.

6.3.2.2   Onset Rate

Onset represents the occurrence of a new sound (melodic or percussive) in a studied signal. For a given musical excerpt, the onset rate reflects the number of onsets per second. In Essentia, the onset detection functions are computed using high frequency content and complex-domain methods [7, 10].

From the box plot of the onset rate feature for four emotions (Fig. 6.8), we can see that the mean value of the onset rate is much higher in excerpts labeled with emotions from the higher quarters of Russell's model. Note density (onsets) is quite similar for

Fig. 6.8 Box plot of the normalized onset rate feature for data set labeled with four emotions e1–e4

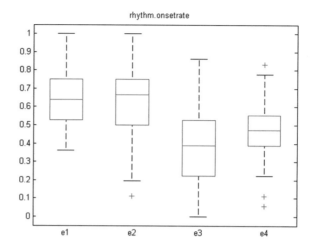

excerpts labeled with e1 and e2, and is much lower for excerpts labeled with e3 and e4. This means that the onset rate feature could be useful for differentiating emotions on the arousal axis.

6.3.3 Tonal Features

6.3.3.1 Mode

Music in West-European culture can generally have two modes: major and minor. This is connected to the structure of the musical scale on the basis of which a composition is built. In the major scale major chords dominate, and in the minor scale minor chords dominate. The major chord differs from the minor chord only by the placement of the middle sound of the chord, but the acoustic sounds of the chords are radically different. The major chord is associated with simplicity, strength, a positive and pleasant sound; while the minor chord evokes negative, sad, and imperfect associations.

Gomes [27] proposed a method for obtaining the mode of a musical excerpt. First, the recorded signal is transformed from time to the spectral domain by FFT. Next, frequencies between 100 Hz and 5000 Hz are used to locate spectral peaks. Frequency deviations of the located spectral peaks are analyzed to find reference frequencies, which are then used to build a Harmonic Pitch Class Profile (HPCP) vector. This vector represents the intensities of the twelve semitone pitch classes (notes from A to G#) found in the musical excerpt. By comparing the HPCP vector to the minor and the major reference key profiles, the major or minor mode is estimated.

If we compare the estimated minor and major mode distributions in music excerpts labeled with four emotions (Fig. 6.9), we notice that we have a large difference

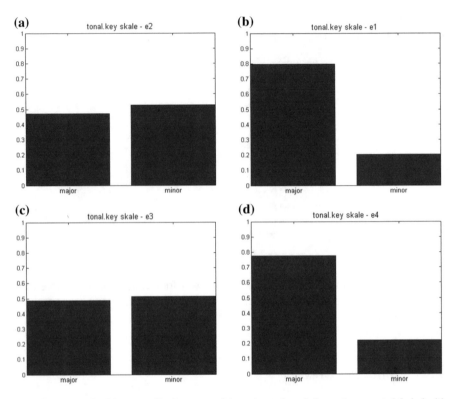

Fig. 6.9 Bar graph of the normalized amount of the estimated mode in music excerpts labeled with four emotions e1–e4

between excerpts that differ by the emotion on the valence axis (Fig. 6.9a ↔ b, c ↔ and d). With musical examples labeled with an emotion placed on positive valence (e1, e4), the estimated major mode is four times more frequent than minor (Fig. 6.9b and d). With examples placed on negative valence, the amount estimated in minor mode is only slightly bigger than major (Fig. 6.9a and c). It is a very interesting result when we compare distributions on the arousal axis, because they are quite similar (Fig. 6.9b ↔ d and a ↔ c). Summarizing these findings, we can state that the mode is very helpful in distinguishing emotions on the valence axis and useless on the arousal axis.

6.3.3.2   Chord Change Rate

The chord change rate is a rough estimator of the number of chord changes per second in a musical excerpt. To count this feature, chord detection should be performed first. Using pitch profile classes, the chord detection algorithm calculates the best matching major or minor chord.

Fig. 6.10 Box plot of normalized chord change rate for data set labeled with four emotions e1–e4

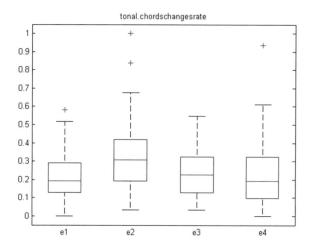

The chord change rate is particularly useful in distinguishing excerpts labeled with emotion e2 from others. From the chord change rate distribution (Fig. 6.10) we see that for excerpts labeled with e2, the median value is higher than for emotions that are not e2.

6.3.3.3 Key Strength

Key strength is estimated from the Harmonic Pitch Class Profile (HPCP) vector. This feature represents the strength of the estimated key.

When we compare key strength distributions for four emotions in music excerpts (Fig. 6.11), we notice that this feature is important for identifying emotions on the

Fig. 6.11 Box plot of normalized key strength for data set labeled with four emotions e1–e4

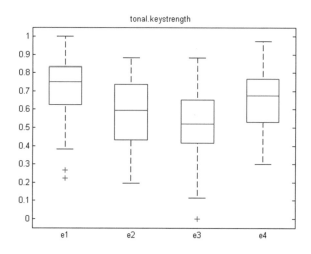

valence axis. Musical samples labeled with positive emotions (e1, e4) are differentiated by higher median values of key strength than samples labeled with negative emotions (e2, e3). The lowest key strength median value is observed for samples with emotion e3.

6.4   Conclusions

We presented some of the most relevant features, showing their potential to individually discriminate between categories. Perfunctory analysis of the value distribution of selected features for audio excerpts labeled with four basic emotions suggests that some features are more useful for detecting emotions on the valence axis while others are more effective for differentiating emotions on the arousal axis.

Tonal features demonstrate better properties for distinguishing emotions on the valence axis (Key Strength, Chord Change Rate, Mode). Rhythm features should potentially be better at differentiating emotions on the arousal axis (Onset Rate, Danceability). Although it is difficult to determine to which axis of the model of emotions timbre features are more predisposed, to change their values between four emotions (Spectral Centroid, Dissonance), timbre features seem to be useful for detecting emotions on both axes simultaneously.

However, a rich set of features along with various statistics obtained from audio analysis tools are kept for automatic emotion detection. Features that are insignificant independently can be important combined with other features. The decision as to selecting the most important features while building an emotion prediction model has been left to the algorithms for feature selection.

Chapter 7
Detection of Four Basic Emotions

7.1 Introduction

One of the most important elements when listening to music is the expressed emotions. The emotions contained in music can alter or deepen the emotional state of the listener. For example, the Funeral March listened to during a funeral deepens the emotional state of the departed's loved ones; while light and relaxing music listened to at home after a hard day's work can restore the listener's good mood. The elements of music that affect the emotions are timbre, dynamics, rhythm, and harmony. Changes in the types of instruments used, the dynamics, rhythm, and harmony change the emotions found in the music.

In this chapter, we study the quality of the constructed music emotion detection classifiers using audio features extracted by two different analysis tools: Essentia [8] and Marsyas [106]. We also decided to study the effect of extracted audio features on the quality of the constructed music emotion detection classifiers. We selected features and found sets of features that were the most useful for detecting individual emotions. We examined the effect of low-level, rhythm and tonal features on the accuracy of the constructed classifiers.

7.2 Related Work

Studies on emotion detection in music are mainly based on two popular approaches: categorical or dimensional. In the dimensional approach, emotions are described as numerical values of valence and arousal. The categorical approach describes emotions with a discrete number of classes – affective adjectives. In this chapter, we used the categorical approach.

© Springer International Publishing AG 2018
J. Grekow, *From Content-Based Music Emotion Recognition to Emotion Maps of Musical Pieces*, Studies in Computational Intelligence 747,
https://doi.org/10.1007/978-3-319-70609-2_7

One of the first papers on categorical emotion detection was a study by Li and Ogihara [57], who trained support vector machines (SVM) to classify music into one of 13 mood categories using a multi-label classification method. A labeled collection consisted of 499 sound files (30-seconds each) from the ambient, classical, fusion, and jazz genres. They used Marsyas to extract the timbral, rhythmic, and pitch features. The achieved accuracy was low, at a level of 45%.

Lu et al. [63] examined emotion detection and emotion tracking using intensity, timbre, and rhythm acoustic features. Emotion categories corresponded to the four quadrants on Thayer's two-dimensional (Energy-Stress) model [103]. To train, Gaussian Mixture Models were used on a set of 800 classical music clips (20 s each). The system of emotion detection achieved an average accuracy of 86%. In addition to emotion detection, emotion tracking through a music piece was presented, which divided the music into several segments.

The problem of multi-label classification of emotions in musical recordings was also presented by Wieczorkowska et al. [113]. The data set contained 875 samples with a length of 30 s each. For classification, the k-nearest neighbors (k-nn) algorithm was used.

In the community of Music Information Retrieval Evaluation eXchange (MIREX) for automatic music mood classification, five mood clusters were used for song categorization [43]. The Audio Mood Classification evaluation task was started for the first time in 2007. The ground truth set consisted of 600 clips (30 second each), with 120 in each mood cluster. The five emotion clusters, which were used by MIREX Audio Mood Classification, have not been frequently used in other music emotion detection works. Hu et al. in [44] indicates that the clusters might not be optimal and noticed some semantic overlap.

A popular emotion set used to categorize emotions in music turned out to be a collection consisting of 4 classes: happy, angry, sad, and relaxed. It corresponds to the four quadrants of the two-dimensional valence-arousal plane, which was used by Laurier in [54], where binary classifiers were constructed for each category. A data set of 1000 songs (30 s each) was divided between 4 categories. Classification accuracy was from 84% to 98%, and was obtained for the SVM algorithm with polynomial and linear kernel.

Four emotion classes (happy, angry, sad, relaxed) were also used in the categorical approach by Song et al. in [100]. The collected ground truth data set consisted of 2904 songs that were labeled with one of the four emotions. The highest accuracy, 53%, was achieved for SVM with polynomial kernel. Song et al. explored the relationship between musical features extracted by MIRtoolbox [53] and emotions. They compared the emotion prediction results for four sets of features: dynamic, rhythm, harmony, and spectral.

7.3 Music Data

In this research, we use four emotion classes: happy, angry, sad, and relaxed. They corresponds to the four quarters of Russell's model [88], which were formed by dividing a plane by two perpendicular axes: arousal and valence. The basic classes of emotions are assigned to the quarters as follows:

- happy – arousal high, valence high – Q1;
- angry – arousal high, valence low – Q2;
- sad – arousal low, valence low – Q3;
- relaxed – arousal low, valence high – Q4.

To conduct the study of emotion detection, we prepared two sets of data. One set was used for building one common classifier for detecting the four emotions, and the other data set for building four binary classifiers of emotion in music. Both data sets consisted of 6-second fragments of different genres of music: classical, jazz, blues, country, disco, hip-hop, metal, pop, reggae, and rock. The tracks were all 22050 Hz mono 16-bit audio files in .wav format. The data set that was used in this experiment consisted of 324 six-second fragments and was described in detail in Chap. 3 Sect. 3.3.

Data annotation was done by five music experts with a university music education. The annotation process of music files with emotion classes was described in Chap. 3 Sect. 3.3.3. The amount of examples in the quarters on the A-V emotion plane is presented in Table 7.1.

To build binary classifiers, we prepared the second training data from the first set, which consisted of four sets of binary data. For example, the data set for binary classifier e1 consisted of 81 files labeled e1 and 81 files labeled not e1 (27 files each from e2, e3, e4). Thus, we obtained four binary data sets (consisting of 81 examples of 'e' and 81 examples of 'not e') for four binary classifiers e1, e2, e3, e4. To make the number of examples uniform in the binary data sets for the four classes, the number of examples labeled e1 was reduced to 81 and the number of those labeled e2 and e3 was reduced to 81.

Table 7.1 Amount of examples in quarters on A-V emotion plane

Basic emotion	Emotion abbreviation	Quarter	Arousal-Valence	Amount of examples
Happy	e1	Q1	High-High	93
Angry	e2	Q2	High-Low	70
Sad	e3	Q3	Low-Low	80
Relaxed	e4	Q4	Low-High	81

7.4   Feature Extraction

For feature extraction, we used Essentia [8] and Marsyas [106], which are tools for audio analysis and audio-based music information retrieval. Marsyas framework was described in Chap. 6 Sect. 6.2.2, and Essentia extractors were described in Chap. 6 Sect. 6.2.1.

The previously prepared, labeled by emotions, music data set served as input data for tools used for feature extraction. The obtained lengths of feature vectors, dependent on the package used, and were as follows: Marsyas – 124 features, and Essentia – 530 features.

7.5   Results

7.5.1   Construction of One Classifier Recognizing Four Emotions

We built classifiers for emotion detection using the following algorithms: J48, RandomForest, BayesNet, K-nn, SMO (SVM). The classification results were calculated using a cross validation evaluation CV-10.

The first important result was that during the construction of the classifier for 2 data sets obtained from Marsyas and Essentia, the highest accuracy among all tested algorithms was obtained for SMO algorithm [79]. SMO was trained using polynomial kernel. The second best algorithm was RandomForest.

The best results we obtained using the feature set from Essentia. The results obtained for SMO algorithm are presented in Table 7.2. The classifier accuracy improved to 64.51% after applying attribute selection (attribute evaluator: Wrapper Subset Evaluator [50], search method: Best First [117]). In Essentia, tonal and rhythm features greatly improve classifier accuracy. These features are not available in Marsyas and thus Essentia obtains better results.

The confusion matrix (Table 7.3), obtained during classifier evaluation, shows that the most recognized emotions were e2 and e4 (F-measure = 0.68), and the next emotion was e1 (F-measure = 0.64). The hardest emotion to recognize was e3 (F-measure = 0.59).

Table 7.2 Accuracy obtained for SMO algorithm

	Essentia (%)	Marsyas (%)
Before attribute selection	62.04	54.01
After attribute selection	**64.51**	58.02

Table 7.3 Confusion matrix for the best result

		Predicted class			
		e1	e2	e3	e4
Actual class	e1	**66**	10	4	13
	e2	21	**42**	5	2
	e3	14	2	**42**	22
	e4	11	0	11	**59**

From the confusion matrix, we can conclude that usually fewer mistakes are made between the top (e1, e2) and bottom (e3, e4) quadrants of Russell's model. At the same time, recognition of emotions on the valence axis (positive-negative) is more difficult.

The most important features, for the detection of four basic emotions, after applying attribute selection were:

- Dissonance (L),
- Melbands Crest (L),
- Melbands Kurtosis (L),
- Melbands Spread (L),
- Spectral Complexity (L),
- Spectral Energy (L),
- Spectral Kurtosis (L),
- Spectral Spread (L),
- Spectral RMS (L),
- Harmonic Pitch Class Profile Entropy (T),
- Tuning Diatonic Strength (T),

where L and T represent feature group abbreviations: low-level (L), tonal (T).

The results were not satisfactory; classifier accuracy was too low (64.51%). It is difficult to build a good classifier that differentiates four emotions equally well.

7.5.2 Construction of Binary Classifiers

To improve emotion detection accuracy, we decided to build specialized binary classifiers for each emotion. A binary classifier algorithm can better analyze data sets for the presence of a given emotion.

During the construction of the binary classifiers, we tested the following algorithms: J48, RandomForest, BayesNet, IBk (K-nn), and SMO (SVM) on the prepared binary data. We calculated the classification results using a cross validation evaluation CV-10. In this experiment, we used features extracted from Essentia, which was selected as the winner in the previous experiment.

Table 7.4 Classifier accuracy for emotions e1, e2, e3, and e4 obtained for SMO

	Classifiers for e1 (%)	Classifiers for e2 (%)	Classifiers for e3 (%)	Classifiers for e4 (%)
Before attribute selection	75.92	80.24	74.07	72.84
After attribute selection	87.04	87.65	82.71	87.04

Table 7.5 Selected features used for building binary classifiers

Classifier	Selected features	Classifier	Selected features
e1	Barkbands Kurtosis (L)	e3	Melbands Crest (L)
	Dissonance (L)		Melbands Kurtosis (L)
	High Frequency Content (L)		Pitch Salience (L)
	Spectral Centroid (L)		Spectral Energy (L)
	Spectral Complexity (L)		Spectral Entropy (L)
	Spectral Entropy (L)		Key Strength (T)
	Spectral Strong Peak (L)		
	Beats Loudness (R)		
	Chords Strength (T)		
	Key Strength (T)		
	Chords Histogram (T)		
e2	Barkbands Flatness (L)	e4	Barkbands Kurtosis (L)
	Melbands Flatness (L)		Barkbands Skewness (L)
	Silence Rate (L)		Barkbands Spread (L)
	Spectral Entropy (L)		Melbands Crest (L)
	Onset Rate (R)		Spectral Complexity (L)
	Chords Strength (T)		Beats Loudness Band Ratio (R)
			Harmonic Pitch Class Profile (T)

Once again, we obtained the best results for SMO algorithm, which are presented in Table 7.4. Accuracy improved (7–15 % points) for all four classifiers after applying attribute selection (attribute evaluator: Wrapper Subset Evaluator, search method: Best First).

The best classifier accuracy was obtained for emotion e2 (87.65%); the results were also high for e1 and e4 (87.04%). Summarizing, accuracy is higher than 80% for all emotions, which is a big improvement of accuracy in comparison with the previous experiment, where we used one classifier recognizing four emotions (64.51%).

Table 7.5 presents the most important features obtained after feature selection (attribute evaluator: Wrapper Subset Evaluator, search method: Best First) for each emotion. Each classifier dedicated to recognizing only one emotion has its own set of features, different from the rest, consisting of a combination of low-level, rhythm, and tonal features. We can notice a domination of low-level features. Features describing

Table 7.6 Classifier accuracy for emotions e1, e2, e3, and e4 obtained for combinations of feature sets

Feature set	Classifiers for e1 (%)	Classifiers for e2 (%)	Classifiers for e3 (%)	Classifiers for e4 (%)
L	79.01	86.42	77.77	85.80
R	76.54	83.33	78.93	77.16
T	75.93	79.63	82.10	77.77
L + R	77.16	89.50	80.25	85.80
L + T	**87.04**	**92.59**	**82.71**	86.42
R + T	**87.04**	83.95	**82.71**	75.30
All (L + R + T)	**87.04**	87.65	**82.71**	**87.04**

spectrum occur in all four sets. Features describing energy in the Barkbands of a spectrum occur in three sets (e1, e2, e4). Features describing energy in the Melbands of a spectrum occur in three sets (e2, e3, e4). Tonal features, Chord Strength, and Key Strength are also important since they are included in two sets each.

7.5.3 Evaluation of Different Combinations of Feature Sets

During this experiment, we evaluated the effect of various combinations of feature sets – low-level (L), rhythm (R), tonal (T) – on classifier accuracy obtained for SMO algorithm (Table 7.6). The best results for each classifier have been marked in bold.

The obtained results indicate that the use of all groups (low-level, rhythm, tonal) of features resulted in the best accuracy or equal with use of 2 groups of features, in most cases (e1, e3, e4). The only exception was classifier e2, where using the set L + T (low-level, tonal) had better results (92.59%) than using all features – accuracy 87.65%.

The use of individual feature sets L, R or T did not have better results than their combinations. Combining feature sets L + T (low-level and tonal features) improved classifier results in the case of all classifiers (e1 e2, e3 and e4). Combining feature sets R + T (rhythm and tonal features) improved classifier results in the case of classifiers e1 and e3.

7.5.4 Emotion Maps

The result of emotion tracking of musical compositions are emotion maps. We used the best obtained classifier for predicting four emotions to analyze musical compositions. The compositions were divided into 6-second segments with a 3/4 overlap.

Fig. 7.1 Emotion map for
the song Let It Be by Paul
McCartney (The Beatles)

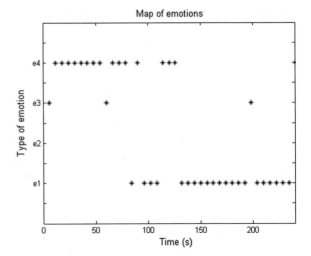

Fig. 7.2 Emotion map for
Piano Sonata No. 8 in C
minor, Op. 13 (Pathetique),
2nd movement, by Ludwig
van Beethoven

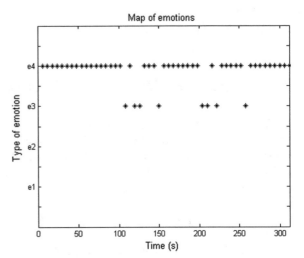

For each segment, features were extracted and classifiers for emotion detection were
used.

From the created emotion maps, we can find out:

- which emotion or emotions are dominant throughout the entire composition,
- how often changes in emotions occur and in which directions,
- which emotions are shaped and developed during the duration of the composition:
  at the beginning, in the middle, and at the end of the piece.

Figures 7.1 and 7.2 show emotion maps of two compositions, one for the song
Let It Be by Paul McCartney (The Beatles) and the second, Piano Sonata No. 8
in C minor, Op. 13 (Pathetique), 2nd movement, by Ludwig van Beethoven. The

horizontal axis shows the time in seconds and the vertical axis the emotion occurring at a given moment.

From the emotion maps of the presented compositions, we can notice their diametrically different emotional character. In the case of Sonata Pathetique, presented in Fig. 7.2, the dominating emotion is e4 (relaxed); and in Let It Be, presented in Fig. 7.1, the dominating emotions are e1 (happy) and e4 (relaxed). Analyzing the development of emotions over time, we notice that in Let It Be there is a different emotion at the beginning and a different one at the end of the composition; in the beginning part of the piece (up until about 90 s.) e4 dominates, and in the end (from 130 s.) e1 dominates, while emotion e3 (sad) occurs sporadically and for a short time (s. 10, 60, 200). In Sonata Pathetique, emotion e4 dominates throughout the entire composition, with short changes in the direction of emotion e3 (s. 120, 200).

The presented emotion maps can have various applications. They can be used to search the database for compositions with a similar or specified distribution of emotions. After extracting parameters describing emotions on a map, we could compare groups of compositions or even compositions by various composers [32].

The emotion maps of compositions using four basic emotions (happy, angry, sad and relaxed) are, however, an oversimplification of the many shades of emotions occurring in music. The detailed distribution of emotions over time of the aforementioned compositions is presented in Chap. 9 Sect. 9.8.1, where we used the Arousal-Valence plane to build emotion maps.

7.6 Conclusions

In this chapter, we presented the detection of four basic emotions in music files. We built a classifier recognizing four basic emotions, but its accuracy was not satisfactory (64%). We then built 4 binary classifiers dedicated to each emotion, with much higher accuracy, from 82% to 87%.

We studied the effect of the extracted audio features on the quality of the constructed music emotion detection classifiers. We obtained information about which features are useful in the detection of particular emotions. The use of all three groups (low-level, rhythm, tonal) of features resulted in the best accuracy or equal with the use of two groups of features, in most cases of binary classifiers.

As a result of emotion tracking of musical compositions, we constructed emotion maps visualizing the distribution of emotions over time. Emotion maps provide new knowledge about the distribution of four emotions in musical compositions and can be used to search for compositions with a specified distribution of emotions, among others.

Chapter 8
Emotion Tracking of Radio Station Broadcasts

8.1 Introduction

The overwhelming number of media outlets is constantly growing. This also applies to radio stations available on the Internet, over satellite and air. On the one hand, the number of opportunities to listen to various radio shows has grown, but on the other, choosing the right station has become more difficult. Music information retrieval helps those people who listen to the radio mainly for the music. This technology is able to make a general detection of the genre, artist, and even emotion.

Listening to music is particularly emotional. People need a variety of emotions, and music is perfectly suited to provide them. Listening to a radio station throughout the day, whether we want it or not, we are affected by the transmitted emotional content. In this paper, we focus on emotional analysis of the music presented by radio stations. During the course of a radio broadcast, these emotions can take on a variety of shades, change several times with varying intensity. This paper presents a method of tracking changing emotions during the course of a radio broadcast. The collected data allowed to determine the dominant emotion in the radio broadcast and construct maps visualizing the distribution of emotions over time.

There are studies focused on facilitating radio station selection from the overwhelming number of radio stations. A method for profiling radio stations was described by Lidy and Rauber [58], who used a technique of Self-organizing Maps to organize the program coverage of radio stations on a two-dimensional map. This approach allows profiling the complete program of a radio station.

A study that combines emotion detection and facilitating radio station selection was presented by Rizk et al. [86], who presented a mobile application that streams music from online radio stations after identifying the user's emotions. The songs from online radio stations were classified into emotion classes based on audio features. The application captured images of the user's face using a smartphone camera and classified them into one of three emotions using a classifier on facial geometric distances and wrinkles.

© Springer International Publishing AG 2018
J. Grekow, *From Content-Based Music Emotion Recognition to Emotion Maps of Musical Pieces*, Studies in Computational Intelligence 747,
https://doi.org/10.1007/978-3-319-70609-2_8

The issue of emotion tracking is not only limited to music. The paper by Mohammad [68] is an interesting extension of the topic; the author investigated the development of emotions in literary texts. Yeh et al. [120] tracked the continuous changes of emotional expressions in Mandarin speech.

8.2   System Construction

The proposed system for tracking emotions in radio station broadcasts is shown in Fig. 8.1. It is composed of collected audio data, a segmentation module, a feature extraction module, classifiers, and a result presentation module.

The recorded radio station broadcasts undergo segmentation, and the obtained fragments are then analyzed in the feature extraction module. The example represented by vectors composed of extracted features then undergo classification by a music/speech classifier. Fragments containing music are additionally classified in terms of emotions. In the last phase, the results are analyzed and visualized in the result presentation module. The music speech classifier and emotion classifier used in this process are trained using features obtained from audio samples.

Fig. 8.1 System construction of emotion tracking in radio station broadcasts

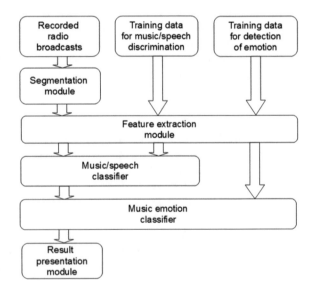

8.3 Music Data

8.3.1 Training Data

To conduct the study of emotion detection of radio stations, we prepared two sets of training data:

1. Data set used for music/speech discrimination;
2. Data set used for the detection of emotion in music.

The set of training data for music/speech discrimination consisted of 128 files, including 64 designated as speech and 64 marked as music. The tracks were all 22050 Hz mono 16-bit audio files in .wav format. The training data were taken from the generally accessible data collection for the purposes of music/speech discrimination from MARSYAS[1] project.

The training data set for emotion detection consisted of 324 six-second fragments of different genres of music: classical, jazz, blues, country, disco, hip-hop, metal, pop, reggae, and rock. The tracks were all 22050 Hz mono 16-bit audio files in .wav format. The data set has been described in detail in Chap. 3 Sect. 3.3. Data annotation was done by five music experts with a university music education. The annotation process of music files with emotion classes has been described in Chap. 3 Sect. 3.3.

In this research, we use four emotion classes corresponding to the four quarters of Russell's model: happy, angry, sad, and relaxed. The amount of examples in the training data set for emotion detection labeled by emotions are presented in Table 8.1.

8.3.2 Recorded Radio Broadcasts

To study changes in emotions, we used recorded broadcasts from 4 selected European radio stations:

- Polish Radio Dwojka (Classical/Culture), recorded on 4.01.2014;
- Polish Radio Trojka (Pop/Rock), recorded on 2.01.2014;
- BBC Radio 3 (Classical), recorded on 25.12.2013;
- ORF OE1 (Information/Culture), recorded on 12.01.2014.

For each station, we recorded 10 h beginning at 10 A.M. and converted the recordings into 22050 Hz mono 16-bit audio files in .wav format. The recorded broadcasts were segmented into 6-second fragments, for example, we obtained 6000 segments from one 10 h broadcast.

---

[1]http://marsyas.info/downloads/datasets.html.

Table 8.1 Amount of examples labeled by emotions

Basic emotion	Emotion abbreviation	Amount of examples
Happy	e1	93
Angry	e2	70
Sad	e3	80
Relaxed	e4	81

8.4 Feature Extraction

For feature extraction for music/speech discrimination, we used the Marsyas framework for audio processing [106], which has been described in detail in Chap. 6 Sect. 6.2.2. For feature extraction for emotion detection, we used the Essentia extractors [8], which have been described in Chap. 6 Sect. 6.2.1.

Essentia has a much richer feature set than Marsyas and is better suited for feature extraction for emotion detection. Marsyas, with its modest feature set, is enough for good music/speech discrimination.

For each 6-second file from the training data, we obtained a representative single feature vector. The obtained vectors were then used for building classifiers and for predicting new instances.

8.5 Construction of Classifiers

8.5.1 Music/Speech Classifier

We built two classifiers, one for music/speech discrimination and the second for emotion detection. During the construction of the classifier for music/speech discrimination, we tested the following algorithms: J48, RandomForest, IBk (K-nn), BayesNet, SMO (SVM). The classification results were calculated using a cross validation evaluation CV-10.

The best accuracy (98%) was achieved using SMO algorithm, which is an implementation of support vector machines (SVM) algorithm (Table 8.2). The confusion matrix for the best music/speech classifier obtained for SMO algorithm is presented in Table 8.3.

Table 8.2 Accuracy and F-measure obtained for tested algorithms

	J48	RandomForest	BayesNet	IBk	SMO
Accuracy (%)	89.84	96.09	95.31	96.09	**98.44**
F-measure	0.89	0.96	0.95	0.96	**0.98**

Table 8.3 Confusion matrix for music/speech classifier obtained for SMO algorithm

		Predicted class	
		Music	Speech
Actual class	Music	**63**	1
	Speech	1	**63**

8.5.2 Classifier for Emotion Detection

We built classifiers for emotion detection using the following algorithms: J48, RandomForest, BayesNet, IBk (K-nn), SMO (SVM). The classification results were calculated using a cross validation evaluation CV-10.

For emotion detection, we used four binary classifiers dedicated to each emotion. The process of building the binary classifiers for emotion detection has been presented in Chap. 7 Sect. 7.5.2. The best classifier accuracy was obtained for emotion e2 (87.65%), but for e1 and e4 the results were also high (87.04%). The lowest classifier accuracy was obtained for emotion e3 (82.71%).

From the data obtained during classifier construction, we can clearly see that music/speech discrimination in audio recordings is a much easier task (98% accuracy) than emotion detection (accuracy from 82 to 87%). The reason behind this is that the audio feature set that can discriminate music from speech is particularly comprehensive. In the case of emotion detection, the feature set is not yet so ideal.

8.6 Results of Emotion Tracking of Radio Stations

During the analysis of the recorded radio broadcasts, we conducted a two-phase classification. The recorded radio program was divided into 6-second segments. For each segment, we extracted a feature vector, which was first used to detect if the given segment is speech or music. If the current segment was music, then we used a second classifier to predict what type of emotion it contained. For feature extraction, file segmentation, use of classifiers to predict new instances, and visualization of results, we wrote a Java application that connected different software products: Marsyas, Essentia, MATLAB and WEKA package.

The percentages of speech, music, and emotion in music obtained during the segment classification of 10-hour broadcasts of four radio stations are presented in Table 8.4. On the basis of these results, radio stations can be compared in two ways: the first is to compare the amount of music and speech in the radio broadcasts, and the second is to compare the occurrence of individual emotions.

Table 8.4 Percentage of speech, music, and emotion in music in 10-hour broadcasts of four radio stations

	PR Dwojka(%)	PR Trojka(%)	BBC Radio 3(%)	ORF OE1(%)
Speech	59.37	73.35	32.25	69.10
Music	40.63	26.65	67.75	30.90
e1	4.78	4.35	2.43	2.48
e2	5.35	14.43	1.00	0.92
e3	20.27	6.02	56.19	22.53
e4	10.23	1.85	8.13	4.97
e1 in music	11.76	16.32	3.58	8.02
e2 in music	13.16	54.14	1.47	2.98
e3 in music	49.89	22.59	82.93	72.91
e4 in music	25.17	6.94	12.00	16.08

8.6.1 Comparison of Radio Stations

The dominant station in the amount of music presented was BBC Radio 3 (67.75%). We noted a similar ratio of speech to music in the broadcasts of PR Trojka and ORF OE1, in both of which speech dominated (73.35% and 69.10%, respectively). A more balanced amount of speech and music was noted on PR Dwojka (59.37% and 40.63%, respectively).

Comparing the content of emotions, we can see that PR Trojka clearly differs from the other radio stations, because the dominant emotion is e2 energetic-negative (54.14%) and e4 calm-positive occurs the least often (6.94%).

We noted a clear similarity between BBC Radio 3 and ORF OE1, where the dominant emotion was e3 calm-negative (82.93% and 72.91%, respectively). Also, the proportions of the other emotions (e1, e2, e4) were similar for these stations. We could say that emotionally these stations are similar, except that considering the speech to music ratio, BBC Radio 3 had much more music.

The dominant emotion for PR Dwojka was e3, which is somewhat similar to BBC Radio 3 and ORF OE1. Compared to the other stations, PR Dwojka had the most (25.17%) e4 calm-positive music.

8.6.2 Emotion Maps of Radio Station Broadcasts

The figures (Figs. 8.2, 8.3, 8.4 and 8.5) present speech and emotion maps for each radio broadcast. Each point on the map is the value obtained from the classification of a 6-second segment. These show which emotions occurred at given hours of the broadcasts.

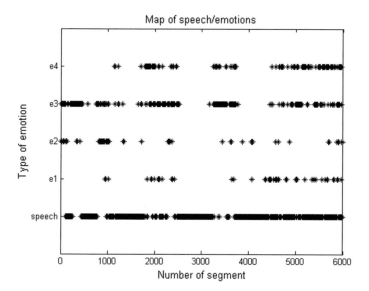

Fig. 8.2 Map of speech and music emotion in PR Dwojka 10 h broadcast

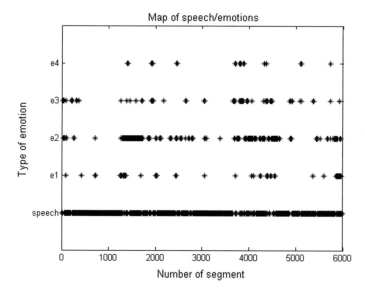

Fig. 8.3 Map of speech and music emotion in PR Trojka 10 h broadcast

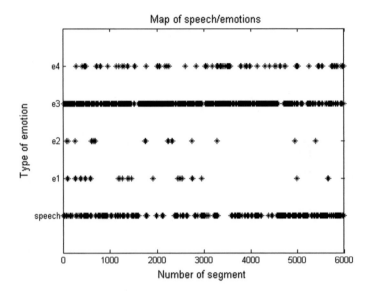

Fig. 8.4 Map of speech and music emotion in BBC Radio 3 10h broadcast

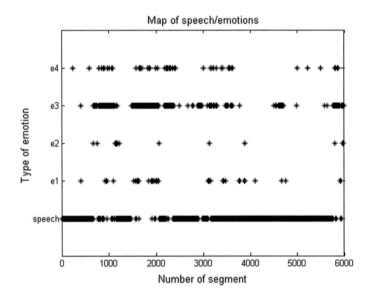

Fig. 8.5 Map of speech and music emotion in ORF OE1 10h broadcast

For PR Dwojka (Fig. 8.2), there are clear musical segments (1500–2500, 2300–3900) during which e3 dominated. At the end of the day (4500–6000), emotion e2 occurs sporadically. It is interesting that e1 and e4 (from right half of Russell's model) did not occur in the morning. For PR Trojka (Fig. 8.3), emotion e4 did not occur in the morning, and e2 and e3 dominated (segments 1200–2800 and 3700–6000). For BBC Radio 3 (Fig. 8.4), we observed almost a complete lack of energetic emotions (e1 and e2) in the afternoon (segments after 3200). For ORF OE1 (Fig. 8.5), e3 dominated up to segment 3600, and then broadcasts without music dominated. The presented analyses of maps of emotions could be developed by examining the quantity of changes of emotions or the distribution of daily emotions.

8.7   Conclusions

This chapter presented an example of a system for the analysis of emotions contained within radio broadcasts. The collected data allowed to determine the dominant emotion in the radio broadcast and present the amount of speech and music. The obtained results provide a new interesting view of the emotional content of radio stations.

A system for the analysis of emotions contained within radio broadcasts could be a helpful tool for people planning radio programs enabling them to consciously plan the emotional distribution in the broadcast music. Another example of applying this system could be an additional tool for radio station searching.

Chapter 9
Music Emotion Maps in the Arousal-Valence Space

9.1 Introduction

Emotions are a dominant element in music, and they are the reason people listen to music so often [81]. Systems searching musical compositions on Internet databases more and more often add an option of selecting emotions to the basic search parameters, such as title, composer, genre, etc. [40, 85].

The emotional content of music is not always constant, and even in classical music or jazz changes often. Analysis of emotions contained in music over time is a very interesting aspect of studying the content of music. It can provide new knowledge on how the composer emotionally shaped the music or why we like some compositions more than others.

9.2 Related Work

Music emotion recognition concentrates on static or dynamic changes over time. Static music emotion recognition uses excerpts from 15 to 30 seconds and omits changes in emotions over time. It assumes the emotion in a given segment does not change. A regression approach and static emotion recognition was presented in [60, 109, 119].

Dynamic music emotion recognition analyzes changes in emotions over time. Methods for detecting emotions using a sliding window are presented in [32, 34, 51, 63, 96, 119]. Deng and Leung [16] proposed multiple dynamic textures to model emotion dynamics over time. To find similar sequence patterns of musical emotions, they used subsequence dynamic time warping for matching emotion dynamics. Aljanaki et al. [3] investigated how well structural segmentation explains emotion segmentation. They evaluated different unsupervised segmentation methods on the

© Springer International Publishing AG 2018
J. Grekow, *From Content-Based Music Emotion Recognition to Emotion Maps of Musical Pieces*, Studies in Computational Intelligence 747, https://doi.org/10.1007/978-3-319-70609-2_9

task of emotion segmentation. Imbrasaite et al. [46] and Schmidt et al. [95] used Continuous Conditional Random Fields for dimensional emotion tracking.

In our study, we used dynamic music emotion recognition with a sliding window. We experimentally selected a segment length of 6 sec. as the shortest period of time after which a music expert can recognize an emotion.

The elements of music that affect the emotions are timbre, dynamics, rhythm, and harmony. One of the most important steps during building a system for automatic emotion detection is feature extraction from audio files. The quality of these features and connecting them with elements of music such as rhythm, harmony, melody and dynamics, shaping a listener's emotional perception of music, have a significant effect on the effectiveness of the built prediction models.

Most papers, however, focus on studying features using a classification model [35, 36, 73, 90, 100]. Music emotion recognition combining standard and melodic features extracted from audio was presented by Panda et al. in [73]. Song et al. [100] explored the relationship between musical features extracted by MIRtoolbox [53] and emotions. They compared the emotion prediction results for four sets of features: dynamic, rhythm, harmony, and spectral. Baume et al. [6] evaluated different types of audio features using a five-dimensional support vector regressor in order to find the combination that produces the best performance.

9.3   Music Data

The data set that was used in this experiment consisted of 324 six-second fragments of different genres of music: classical, jazz, blues, country, disco, hip-hop, metal, pop, reggae, and rock. The tracks were all 22050 Hz mono 16-bit audio files in .wav format. The data set has been described in detail in Chap. 3 Sect. 3.3.

During the annotation of music samples, we used Russell's two-dimensional valence-arousal (V-A) model to measure emotions in music [88]. The model consists of two independent dimensions of valence (horizontal axis) and arousal (vertical axis). The annotation process of music files has been described in Chap. 3 Sect. 3.3.3. The amount of examples in the quarters on the A-V emotion plane is presented in Table 9.1.

Table 9.1   Amount of examples in quarters on A-V emotion plane

Quarter abbreviation	Arousal-Valence	Amount of examples
Q1	High-High	93
Q2	High-Low	70
Q3	Low-Low	80
Q4	Low-High	81

9.4 Feature Extraction

For feature extraction, we used Essentia [8] and Marsyas [106], which are tools for audio analysis and audio-based music information retrieval. Marsyas framework has been described in Chap.6 Sect. 6.2.2. Essentia extractors have been described in Chap. 6 Sect. 6.2.1.

The previously prepared, labeled by A-V values, music data set served as input data for tools used for feature extraction. The obtained lengths of feature vectors, dependent on the package used, were as follows: Marsyas—124 features, and Essentia—530 features.

9.5 Regressor Training

We built regressors for predicting arousal and valence using the WEKA package [114]. For training and testing, the following regression algorithms were used: SMOreg, REPTree, M5P. SMOreg algorithm [99] implements the support vector machine for regression. REPTree algorithm [41] builds a regression tree using variance and prunes it using reduced-error pruning. M5P implements base routines for generating M5 Model trees and rules [83, 110].

Before constructing regressors, arousal and valence annotations were scaled between $[-0.5, 0.5]$. We evaluated the performance of regression using the tenfold cross validation technique (CV-10).

The highest values for determination coefficient (R^2) were obtained using SMOreg (implementation of the support vector machine for regression) [99]. After applying attribute selection (attribute evaluator: Wrapper Subset Evaluator [50], search method: Best First [117]), we obtained $R^2 = 0.79$, for arousal and $R^2 = 0.58$ for valence. Mean absolute error reached values $MAE = 0.09$ for arousal and $MAE = 0.10$ for valence (Table 9.2).

Predicting arousal is a much easier task for regressors than valence in both cases of extracted features (Essentia, Marsays) and values predicted for arousal are more precise. R^2 for arousal were comparable (0.79 and 0.73), but features which describe valence were much better using Essentia for audio analysis. The obtained $R^2 = 0.58$ for valence are much higher than $R^2 = 0.25$ using Marsyas features. In Essentia,

Table 9.2 R^2 and MAE obtained for SMOreg

| | Essentia | | | | Marsyas | | | |
| | Arousal | | Valence | | Arousal | | Valence | |
	R^2	MAE	R^2	MAE	R^2	MAE	R^2	MAE
Before attribute selection	0.48	0.18	0.27	0.17	0.63	0.13	0.15	0.16
After attribute selection	**0.79**	**0.09**	**0.58**	**0.10**	0.73	0.11	0.25	0.14

tonal and rhythm features greatly improve prediction of valence. These features are not available in Marsyas and thus Essentia obtains better results.

One can notice the significant role of the attribute selection phase, which generally improves prediction results. Marsyas features before attribute selection outperform Essentia features for arousal detection. $R^2 = 0.63$ and $MAE = 0.13$ by Marsyas are better results than $R^2 = 0.48$ and $MAE = 0.18$ by Essentia. However, after selecting the most important attribute, Essentia turns out to be the winner with $R^2 = 0.79$ and $MAE = 0.09$.

9.6   Evaluation of Different Combinations of Feature Sets

During this experiment, we evaluated the effect of various combinations of Essentia feature sets—low-level (L), rhythm (R), tonal (T)—on the performance obtained for SMOreg algorithm. We evaluated the performance of regression using the tenfold cross validation technique (CV-10). We also used attribute selection with Wrapper Subset Evaluator and search method Best First.

The obtained results, presented in Table 9.3, indicate that the use of all groups (low-level, rhythm, tonal) of features resulted in the best performance or equal to best performance by combining feature sets. The best results have been marked in bold. Detection of arousal using the set L+R (low-level, rhythm features) has equal results as using all groups. Detection of valence using the set L+T (low-level, tonal features) has only little worse results than using all groups.

The use of individual feature sets L, R or T did not achieve better results than their combinations. Worse results were obtained when using only tonal features for arousal ($R^2 = 0.53$ and $MAE = 0.14$) and only rhythm features for valence ($R^2 = 0.15$ and $MAE = 0.15$).

Combining feature sets L+R (low-level and rhythm features) improved regressor results in the case of arousal. Combining feature sets L+T (low-level and tonal features) improved regressor results in the case of valence.

Table 9.3 R^2 and MAE for arousal and valence obtained for combinations of feature sets

Feature set	Arousal		Valence	
	R^2	MAE	R^2	MAE
L	0.74	0.10	0.49	0.12
R	0.68	0.11	0.15	0.15
T	0.53	0.14	0.48	0.12
L+R	**0.79**	**0.09**	0.40	0.12
L+T	0.74	0.10	**0.56**	**0.10**
R+T	0.74	0.11	0.52	0.11
All (L+R+T)	**0.79**	**0.09**	**0.58**	**0.10**

In summary, we can conclude that low-level features are very important in the prediction of both arousal and valence. Additionally, rhythm features are important for arousal detection, and tonal features help a lot for detecting valence. The use of only individual feature sets L, R or T does not give good results.

9.7 Selected Features Dedicated to the Detection of Arousal and Valence

Table 9.4 presents 2 sets of selected features, which using the SMOreg algorithm obtained the best performance by detecting arousal (Sect. 9.6). Features marked in bold are in both groups. Notice that after adding tonal features T to group L+R, some of the features were replaced by others and some remained without changes. Features found in both groups seem to be particularly useful for detecting arousal. Different statistics from spectrum and mel bands turned out to be especially useful: Spectral Energy, Entropy, Flux, Rolloff, Skewness, and Melbands Crest, Kurtosis. Also, three rhythm features belong to the group of more important features because both sets contain: Danceability, Onset Rate, Beats Loudness Band Ratio.

Table 9.5 presents 2 sets of selected features, which using the SMOreg algorithm obtained the best performance by detecting valence (Sect. 9.6). Particularly important low-level features, found in both groups, were: Spectral Energy and Zero Crossing Rate, as well as Mel Frequency Cepstrum Coefficients (MFCC) and Gammatone Feature Cepstrum Coefficients (GFCC). Particularly important tonal features, which describe key, chords and tonality of a musical excerpt were: Chords Strength, Harmonic Pitch Class Profile Entropy, Key Strength.

Comparing the sets of features dedicated to arousal (Table 9.4) and valence (Table 9.5), we notice that there are much more statistics from spectrum and mel bands in the arousal set than in the valence set. MFCC and GFCC were useful for detecting valence and were not taken into account for arousal detection.

Features that turned out to be universal, useful for detecting both arousal and valence, by using all features (L+R+T), are:

- Melbands Kurtosis (L),
- Melbands Skewness (L),
- Spectral Energy (L),
- Beats Loudness Band Ratio (R),
- Chords Strength (T),
- Harmonic Pitch Class Profile (HPCP) Entropy (T),
- Key Strength (T),
- Chords Histogram (T).

Table 9.4 Selected features used for building the arousal regressor

Features from set L+R+T	Features from set L+R
Average Loudness (L)	Barkbands Kurtosis (L)
Barkbands Spread (L)	Dissonance (L)
Melbands Crest (L)	Erbbands Flatness (L)
Melbands Flatness (L)	Erbbands Skewness (L)
Melbands Kurtosis (L)	**Melbands Crest (L)**
Melbands Skewness (L)	**Melbands Kurtosis (L)**
Melbands Spread (L)	Silence Rate (L)
Spectral Energy (L)	**Spectral Energy (L)**
Spectral Entropy (L)	**Spectral Entropy (L)**
Spectral Flux (L)	**Spectral Flux (L)**
Spectral Kurtosis (L)	**Spectral Rolloff (L)**
Spectral Rolloff (L)	**Spectral Skewness (L)**
Spectral Skewness (L)	Beats Count (R)
Beats Per Minute (BPM) Histogram (R)	Beats Loudness (R)
BPM of the Most Salient Tempo (R)	**Danceability (R)**
Danceability (R)	**Onset Rate (R)**
Onset Rate (R)	**Beats Loudness Band Ratio (R)**
Beats Loudness Band Ratio (R)	
Chords Strength (T)	
Harmonic Pitch Class Profile Entropy (T)	
Key Strength (T)	
Chords Histogram (T)	

Table 9.5 Selected features used for building the valence regressor

Features from set L+R+T	Features from set L+T
High Frequency Content (L)	Melbands Crest (L)
Melbands Kurtosis (L)	Melbands Spread (L)
Melbands Skewness (L)	Pitch Salience (L)
Spectral Energy (L)	Silence Rate (L)
Zero Crossing Rate (L)	Spectral Centroid (L)
GFCC (L)	Spectral Energy (L)
MFCC (L)	Spectral Spread (L)
Beats Loudness (R)	**Zero Crossing Rate (L)**
Onset Rate (R)	**GFCC (L)**
Beats Loudness Band Ratio (R)	**MFCC (L)**
Chords Strength (T)	**Chords Strength (T)**
HPCP Entropy (T)	**HPCP Entropy (T)**
Key Strength (T)	**Key Strength (T)**
Chords Histogram (T)	Key Scale (T)

9.8 Emotion Maps

The result of emotion tracking are emotion maps. We used the best obtained models for predicting arousal and valence to analyze musical compositions. The compositions were divided into 6-second segments with a 3/4 overlap. For each segment, features were extracted and models for arousal and valence were used.

The predicted values are presented in the figures in the form of emotion maps. For each musical composition, the obtained data was presented in 4 different ways:

1. Arousal-Valence over time;
2. Arousal-Valence map;
3. Arousal over time;
4. Valence over time.

Simultaneous observation of the same data in 4 different projections enabled us to accurately track changes in valence and arousal over time, such as tracking the location of a prediction on the A-V emotion plane.

9.8.1 Emotion Maps of Two Compositions

Figures 9.1 and 9.2 show emotion maps of two compositions, one for the song Let It Be by Paul McCartney (The Beatles) and the second, Piano Sonata No. 8 in C minor, Op. 13 (Pathetique), 2nd movement, by Ludwig van Beethoven.

Emotion maps present two different emotional aspects of these compositions. The first significant difference is distribution on the quarters of the Arousal-Valence map. In Let It Be (Fig. 9.1b), the emotions of quadrants Q4 and Q1 (high valence and low-high arousal) dominate. In Sonata Pathetique (Fig. 9.2b), the emotions of quarter Q4 (low arousal and low valence) dominate with an incidental emergence of emotions of quarter Q3 (low arousal and low valence).

Another noticeable difference is the distribution of arousal over time. Arousal in Let It Be (Fig. 9.1c) has a rising tendency over time of the entire song, and varies from low to high. In Sonata Pathetique (Fig. 9.2c), in the first half (s. 0–160) arousal has very low values, and in the second half (s. 160–310) arousal increases incidentally but remains in the low value range.

The third noticeable difference is the distribution of valence over time. Valence in Let It Be (Fig. 9.1d) remains in the high (positive) range with small fluctuations, but it is always positive. In Sonata Pathetique (Fig. 9.2d), valence, for the most part, remains in the high range but it also has several declines (s. 90, 110, 305), which makes valence more diverse.

Arousal and valence over time were dependent on the music content. Even in a short fragment of music, these values varied significantly. From the course of arousal and valence, it appears that Let It Be is a song of a decisively positive nature with a clear increase in arousal over time, while Sonata Pathetique is mostly calm and predominantly positive.

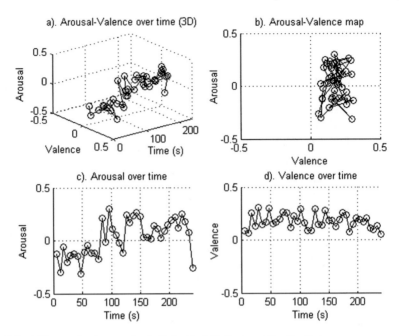

Fig. 9.1 A-V maps for the song Let It Be by Paul McCartney (The Beatles)

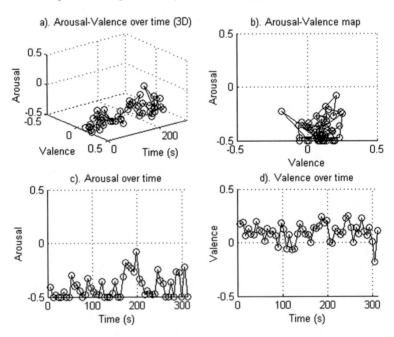

Fig. 9.2 A-V maps for Piano Sonata No. 8 in C minor, Op. 13 (Pathetique), 2nd movement, by Ludwig van Beethoven

9.8.2 Features Describing Emotion Maps

To analyze and compare changes in arousal and valence over time (time series), we proposed the following parameters:

1. *Mean value of arousal*;
2. *Mean value of valence*;
3. *Standard deviation of arousal*;
4. *Standard deviation of valence*;
5. *Mean of derivative of arousal*;
6. *Mean of derivative of valence*;
7. *Standard deviation of derivative of arousal*;
8. *Standard deviation of derivative of valence*;
9. *Quantity of changing sign of arousal QCA*—describes how often arousal changes between top and bottom quarters of the A-V emotion model;
10. *Quantity of changing sign of valence QCV*—describes how often valence changes between left and right quarters of the A-V emotion model;
11. *QCE*—is the sum of *QCA* and *QCV*;
12. *Percentage representation of emotion in 4 quarters* (4 parameters).

Analysis of the distribution of emotions over time gives a much more accurate view of the emotional structure of a musical composition. It provides not only information on which emotions are dominant in a composition, but also how often they change, and their tendency. The presented list of features is not closed, we will search for additional features in the future.

9.8.3 Comparison of Musical Compositions

Another experiment was to compare selected well-known Ludwig van Beethoven's Sonatas with several of the most famous songs by The Beatles. We used nine musical compositions from each group for the comparison (Table 9.6). This experiment did not aim to compare all the works of Beethoven and The Beatles, but only to find the rules and most important features distinguishing these 2 groups.

Each sample was segmented and arousal and valence were detected. Then, 15 features, which were presented in the previous section, were calculated for each sample. We used the PART algorithm [24] from the WEKA package [114] to find the decision-making rules differentiating the two groups.

It turned out that the most distinguishing feature for these two groups of musical compositions was the *Standard deviation of valence*. It was significantly smaller in The Beatles' songs than in Beethoven's compositions (Fig. 9.3). *Standard deviation of valence* reflects how big deviations were from the mean. The results show that in Beethoven's compositions valence values were much more varied than in the songs of The Beatles.

Table 9.6 List of musical compositions

L. v. Beethoven's Sonatas	The Beatles
Sonata Appassionata, part 1	Hey Jude
Sonata Appassionata, part 2	P.S. I Love You
Sonata Appassionata, part 3	While My Guitar Gently Weeps
Sonata Waldstein, part 1	I'll Follow The Sun
Sonata Waldstein, part 2	It's Only Love
Sonata Waldstein, part 3	Yesterday
Sonata Pathetique, part 1	Michelle
Sonata Pathetique, part 2	Girl
Sonata Pathetique, part 3	Let It Be

Fig. 9.3 Box plot of *Standard deviation of valence* in The Beatles' and in Beethoven's compositions

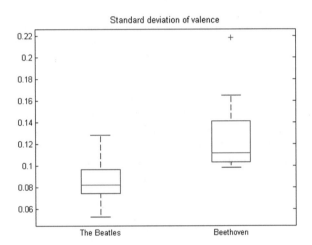

To find another significant feature in the next stage, we removed the characteristic that we found previously (*Standard deviation of valence*) from the data set. Another significant feature was *Standard deviation of arousal*. In Beethoven's compositions, the values of the *Standard deviation of arousal* were much greater than in the Beatles' songs (Fig. 9.4). This proves the compositions have a greater diversity of tempo and volume.

In the next analogous stage, the feature we found was *Standard deviation of derivative of arousal*. It reflects the magnitude of changes in arousal between the studied segments. We found higher values of *Standard deviation of derivative of arousal* in Beethoven's compositions (Fig. 9.5).

An example of a feature that is unsuitable for differentiating between two examined groups of compositions is presented in Fig. 9.6. Overlapping values of the feature *Percentage representation of emotion e4*, obtained for compositions by The Beatles and Beethoven, cause that the usefulness of this feature to differentiate the way emotions are shaped in the studied groups is small.

Fig. 9.4 Box plot of *Standard deviation of arousal* in The Beatles' and in Beethoven's compositions

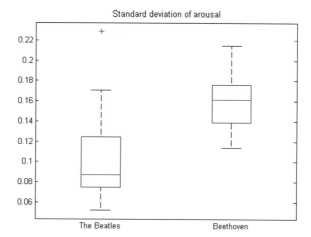

Fig. 9.5 Box plot of *Standard deviation of derivative of arousal* in The Beatles' and in Beethoven's compositions

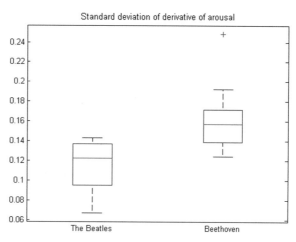

Fig. 9.6 Box plot of *Percentage representation of emotion e4 (relaxed)* in The Beatles' and in Beethoven's compositions

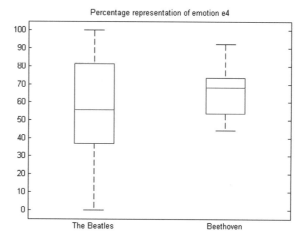

The interesting thing is that in the group of the most important distinguishing features we did not find features describing the emotion type (*Mean value of arousal, Mean value of valence*, or *Percentage representation of emotion in 4 quarters*). This is confirmed by the fact that we cannot assign common emotions to the different sample groups (Beethoven, The Beatles); in all groups, we have emotions from the four quadrants of the emotion model.

We can conclude that features that better distinguish between the two groups of compositions were features pertaining to changes in emotions and their distribution in the musical compositions.

9.9   Conclusions

In this chapter, we presented the detection of emotions as a problem of regression. The result of applying regressors are emotion maps of the musical compositions. Conducting experiments required the construction of regressors, attribute selection, and analysis of selected musical compositions.

Emotion maps provide new knowledge about the distribution of emotions in musical compositions, and knowledge that had only been available to music experts until this point. The proposed parameters describing emotions can be used in the construction of a system that can search for songs with similar emotions. They describe in more detail the distribution of emotions, their evolution, frequency of changes, etc.

In this chapter, we also studied the usefulness of audio features during emotion detection. Different feature sets were used to test the performance of built regression models intended to detect arousal and valence. We examined the influence of different feature sets—low-level, rhythm, tonal, and their combination—on arousal and valence prediction. The use of a combination of different types of features significantly improved the results compared with using just one group of features. We found and presented features particularly dedicated to the detection of arousal and valence separately, as well as features useful in both cases. We can conclude that low-level features are very important in the prediction of both arousal and valence. Additionally, rhythm features are important for arousal detection, and tonal features help a lot for detecting valence.

The obtained results confirm the point of creating new features of middle and higher levels that describe elements of music such as rhythm, harmony, melody, and dynamics shaping a listener's emotional perception of music. These features can have an affect on improving the effectiveness of automatic emotion detection in music files.

Chapter 10
Comparative Analysis of Musical Performances by Using Emotion Tracking on the Arousal-Valence Plane

10.1 Introduction

Musical compositions differ not only in their musical content, but also their emotional message. Even the same composition, based on one musical notation, can be performed differently, with each performance differing in the emotional content. Performing a piece written by a composer, a performer, musician, artist gives it its own shape—interpretation. We can like some performances more than others.

Emotions are increasingly added to basic search parameters, such as title, composer, genre, etc., of systems searching musical compositions on Internet databases. Finding pieces with a similar emotional distribution throughout the same composition is an option that further extends the capabilities of search systems.

In this paper, we present a computer system that enables finding which performances of the same composition are closer to each other and which are quite distant in terms of shaping emotions over time. We analyzed 6 musical works, of which there were 5 different versions.

10.2 Related Work

For comparative analysis of musical performances, we used the dimensional approach of dynamic music emotion recognition. Emotion recognition was treated as a regression problem. We used a 2D emotion model proposed by Russell [88], where the dimensions are represented by arousal and valence. It was used in many works used in music emotion recognition [96, 119]. Dynamic music emotion recognition analyzes changes in emotions over time. Methods for detecting emotion using a sliding window are presented in [38, 51, 63, 96, 119].

Comparisons of multiple performances of the same piece often focused on piano performances [26, 93]. Tempo and loudness information were the most popular

© Springer International Publishing AG 2018
J. Grekow, *From Content-Based Music Emotion Recognition to Emotion Maps of Musical Pieces*, Studies in Computational Intelligence 747,
https://doi.org/10.1007/978-3-319-70609-2_10

characteristics used for performance analysis. They were used to calculate correlations between performances in [93, 94]. In the study [26], tempo and loudness derived from audio recordings were segmented into musical phrases, and then clustering was used to find individual features of the pianists' performances. Four selected computational models of expressive music performance were reviewed in [112]. In addition, research on formal characterization of individual performance style, like performance trajectories and performance alphabets, was presented. A method to compare orchestra performances by examining a visual spectrogram characteristic was proposed in [59]. Principal component analysis on synchronized performance fragments was applied to localize areas of cross-performance variation in time and frequency. A connection between music performances and emotion was presented in [9], where a computer program (Director Musices) was used to produce performances with varying emotional expression. The program used a set of rules characteristic for each emotion (fear, anger, happiness, sadness, solemnity, tenderness), which were used to modify such parameters of MIDI files as tempo, sound level, articulation, tone onsets and delays.

10.3   System Construction

The proposed system for comparative analysis of musical performances using emotion tracking is shown in Fig. 10.1. It is composed of collected music training data, segmentation, feature extraction, regressors, aligning, and a result presentation module.

The input data are different performances of the same composition, which underwent segmentation. After the feature extraction process, the prediction of arousal and valence occurs for subsequent segments, and previously trained regressors are used for prediction. In the next phase, the valence and arousal values are aligned in the aligning module, which causes that the same musical fragments of different performances are compared. The obtained results are sent to the result presentation module, where the course of arousal and valence over time is presented, scape plots are constructed, and parameters indicating the most alike compositions are calculated.

10.4   Music Data for Regressor Training

The data set that was used in this experiment consisted of 324 six-second fragments and has been described in detail in Chap. 3 Sect. 3.3. During the annotation of music samples, we used Russell's two-dimensional valence-arousal (V-A) model to measure emotions in music [88], which consists of two independent dimensions of valence (horizontal axis) and arousal (vertical axis). The annotation process of music files

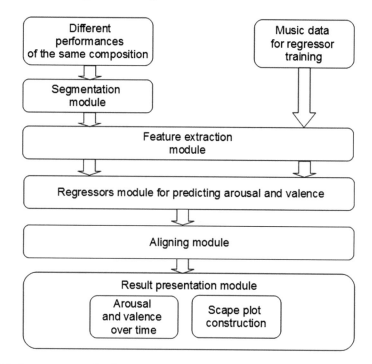

Fig. 10.1   System construction for comparative analysis of musical performances by using emotion tracking

Table 10.1   Amount of examples in quarters on the A-V emotion plane

Quarter abbreviation	Arousal-Valence	Amount of examples
Q1	High-High	93
Q2	High-Low	70
Q3	Low-Low	80
Q4	Low-High	81

has been described in Chap. 3 Sect. 3.3.3. The amount of examples in the quarters on the A-V emotion plane is presented in Table 10.1.

The previously prepared, labeled by A-V values, music data set served as input data for the tool used for feature extraction. For feature extraction, we used a tool for audio analysis Essentia [8]. The obtained by Essentia length of feature vector was 530 features.

10.5   Regressor Training

We built regressors for predicting arousal and valence using the WEKA package [114]. For training and testing, the following regression algorithms were used: SMOreg, REPTree, M5P. Before constructing regressors, arousal and valence annotations were scaled between $[-0.5, 0.5]$.

We evaluated the performance of regression using the tenfold cross validation technique (CV-10). The whole data set was randomly divided into ten parts, nine of them for training and the remaining one for testing. The learning procedure was executed a total of 10 times on different training sets. Finally, the 10 error estimates were averaged to yield an overall error estimate.

The highest values for determination coefficient (R^2) were obtained using SMOreg (implementation of the support vector machine for regression). After applying attribute selection (attribute evaluator: Wrapper Subset Evaluator, search method: Best First), we obtained $R^2 = 0.79$, for arousal and $R^2 = 0.58$ for valence. Mean absolute error reached values $MAE = 0.09$ for arousal and $MAE = 0.10$ for valence. Predicting arousal is a much easier task for regressors than valence and the values predicted for arousal are more precise.

10.6   Aligning of Audio Recordings

Our task was to compare different performances of the same composition using emotional distribution. Because the musical performances are played at varying tempos, with various accelerations and decelerations, an alignment of audio recordings is necessary to compare two performances. This enables comparing the same fragments of different renditions. Without doing an alignment to compare performances second to second, we would be comparing fragments of varying content. Just adjusting the time of different renditions, for example through stretching or compression of time, will not synchronize the performances in terms of music content. Only an exact alignment of the recordings, note by note, guarantees that we are comparing the same fragments.

We used MATCH [18], a toolkit for accurate automatic alignment of different renditions of the same piece of music. MATCH is based on a dynamic time warping algorithm (DTW), which is a technique for aligning time series and has been well known and used in the speech recognition community [84]. Frames of audio input data are represented by positive spectral difference vectors, which emphasize note onsets in the alignment process. They are used in the DTW algorithm's match cost function, which uses an Euclidean metric. The path returned by the DTW algorithm, as result of alignment of two audio files, is used to find the location of the same musical fragment in both files.

Figures 10.2 and 10.3 present, in waveform images, the beginnings of three different renditions of the same composition (Prelude in C major, Op.28, No.1 by Frédéric

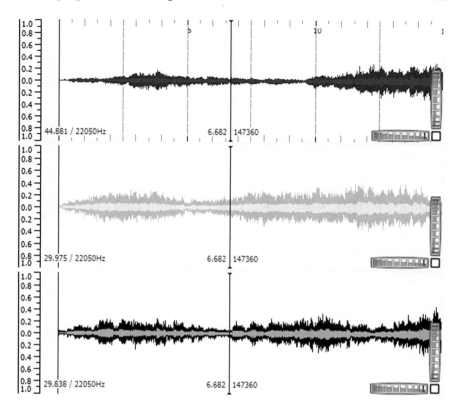

Fig. 10.2 Waveform images of three different music performances of Prelude in C major, Op.28, No.1 by Frédéric Chopin before alignment

Chopin) before and after alignment. Before alignment (Fig. 10.2), the compositions are placed one after the other and the vertical line indicates the time from the beginning of the composition, but these are different fragments in terms of music content. After alignment (Fig. 10.3), the vertical line indicates the same fragment in different performances. We notice the varying locations of the same motif from the beginning of the composition depending on the rendition, which is connected to the differing tempos played by different performers. The top first recording is a reference recording and the remaining pieces are compared to it. To present the waveform images of audio files and to visualize the alignment results, Sonic Visualiser [12] with installed MATCH Vamp Plugin was used.

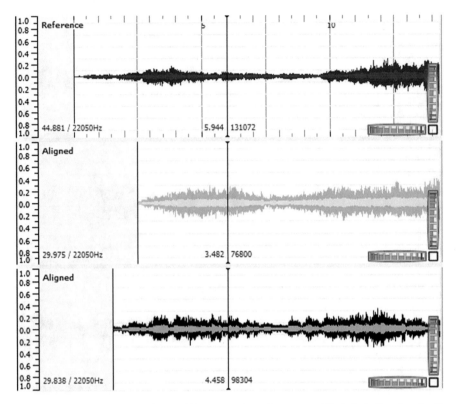

Fig. 10.3 Waveform images of three different music performances of Prelude in C major, Op.28, No.1 by Frédéric Chopin after alignment

10.7   Analyzed Performances

The collection of analyzed performances consisted of the following compositions by Frédéric Chopin (1810–1849):

- Prelude in C major, Op.28, No.1;
- Prelude in D major, Op.28, No.5;
- Prelude in F minor, Op.28, No.18;
- Prelude in C minor, Op.28, No.20 (the first 8 bars).

All the analyzed Chopin performances were audio recordings played by 5 famous pianists:

- Artur Rubinstein recorded in 1946;
- Emil Gilels recorded in 1953;
- Grigory Sokolov recorded in 1990;
- Martha Argerich recorded in 1997;
- Rafał Blechacz recorded in 2007.

Additionally, we analyzed approximately 1 min long beginnings of 2 symphonies by Ludwig van Beethoven (1770–1827): Symphony No.5 in C minor, Op.67 (the first 58 bars) performed by:

- NBC Symphony Orchestra conducted by Arturo Toscanini recorded in 1953;
- Berliner Philharmoniker conducted by Herbert von Karajan recorded in 1982;
- London Philharmonic Orchestra conducted by Horst Stein recorded in 1963;
- New York Philharmonic Orchestra conducted by Leonard Bernstein recorded in 1963;
- Philharmonia Orchestra conducted by Otto Klemperer recorded in 1959;

and Symphony No.3 in E-flat major, Op.55 'Eroica' (the first 46 bars) performed by:

- NBC Symphony Orchestra conducted by Arturo Toscanini recorded in 1953;
- Berliner Philharmoniker conducted by Herbert von Karajan recorded in 1964;
- La Capella Reial de Catalunya conducted by Jordi Saval recorded in 1994;
- New York Philharmonic Orchestra conducted by Leonard Bernstein recorded in 1967;
- Philharmonia Orchestra conducted by Otto Klemperer recorded in 1959.

We analyzed 6 musical works, with 5 different performances of each. Detailed results are available on the web.[1]

10.8 Results

We used the best obtained regression models for predicting arousal and valence of the musical performances, which were divided into 6 s segments with a 3/4 overlap. For each segment, features were extracted and models for arousal and valence were used. As a result, we obtained arousal and valence values for every 1.5 s of a musical piece.

10.8.1 Influence of Recording Alignment on Emotion Values of the Compared Segments

During the comparison of different performances of the same composition, it is very important to compare the same musical fragment of these performances. Thus, we implemented a module for the alignment of audio recordings in our system. We used MATCH [18], a toolkit for accurate automatic alignment of different renditions of the same piece of music. The influence of the alignment process on the obtained

---

[1] http://aragorn.pb.bialystok.pl/~grekowj/HomePage/Performances.

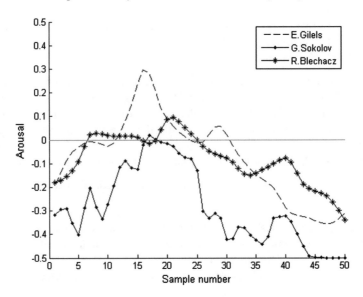

Fig. 10.4 Arousal over time for three different music performances before alignment

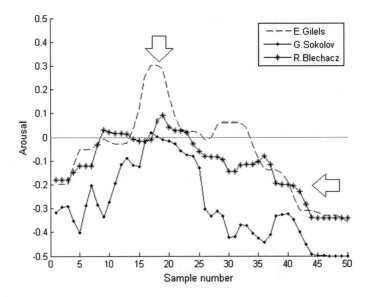

Fig. 10.5 Arousal over time for three different music performances after alignment

values is presented in Figs. 10.4 and 10.5, where arousal values over time before and after alignment are presented (Prelude in C major, Op.28, No.1 by F. Chopin).

Observation of the course of arousal in the performances shows that there was clearly a similar increase in the value for samples 15–23 in the three performances (Fig. 10.4), but it was not well synchronized. The same fragment after alignment (samples 15–20) shows better synchronization. The three performances show their maximum in samples 17–18 (Fig. 10.5). The alignment process is also well illustrated when we compare decreasing arousal values in the performance by R. Blechacz before (samples 40–50, Fig. 10.4) and after alignment (samples 37–50, Fig. 10.5). After alignment, the decreasing arousal values of the performance by R. Blechacz align well to the decreasing arousal values of the performance by E. Gilels.

10.8.2 Performances and Arousal over Time

Due to the limited nature of our research, we have restricted the analysis and presentation of results to 5 different performances of Prelude in C major, Op.28, No.1 by Frédéric Chopin.

Observation of the course of arousal in the performances (Fig. 10.6) shows that the performance by G. Sokolov had significantly lower values than the remaining pieces. The reason for this is that the performance was played at a slower pace and lower sound intensity. Between samples 15 and 20, there was a clear rise in arousal for all performances, but the performance by E. Gilels achieved maximum (Arousal = 0.3). Also, the performance by E. Gilels is the most dynamically aroused in this fragment. It is not always easy to detect on the graph which performances are similar. However, we can notice a convergence of lines between A. Rubinstein and G. Sokolov (samples 30–50), and between R. Blechacz and E. Gilels (samples 37–50).

To compare the musical performances, we used the Pearson correlation coefficient r, which was calculated for each pair of performances (Table 10.2). Each performance is represented by a sequence of arousal values. The correlation coefficient ranges from 1 to -1. Value 1 means perfectly correlated sequences, 0 there is no correlation, and -1 that the sequences are perfectly correlated negatively. A correlation coefficient between two of the same performances is at maximum equal to 1 and was not taken into account, i.e. $r(X, X) = 1$.

Pearson's correlation coefficient r between two sequences $(x_1, x_2, ..., x_n)$ and $(y_1, y_2, ..., y_n)$ of the same length n is defined in Eq. 10.1.

$$r(x, y) = \frac{\sum_{i=1}^{n}(x_i - \bar{x})(y_i - \bar{y})}{\sqrt{\sum_{i=1}^{n}(x_i - \bar{x})^2 \sum_{i=1}^{n}(y_i - \bar{y})^2}} \tag{10.1}$$

where x_i and y_i are values of elements in sequences, $\bar{x}$ and $\bar{y}$ are mean values of each sequence.

Comparing the musical performances in terms of arousal (Table 10.2), we see that the performance by A. Rubinstein was most similar to G. Sokolov ($r = 0.80$). G. Sokolov was most similar to A. Rubinstein ($r = 0.80$) and R. Blechacz ($r = 0.83$). The performances by R. Blechacz were similar to those by E. Gilels ($r = 0.87$) and G. Sokolov ($r = 0.83$). M. Argerich's performance is the least similar to the rest, although it is closest to E. Gilels ($r = 0.74$) and R. Blechacz ($r = 0.75$). From all

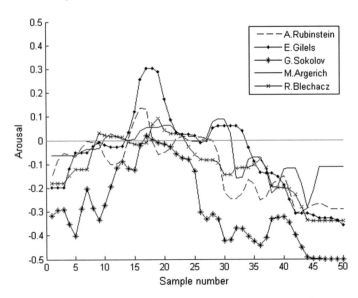

Fig. 10.6 Arousal over time for 5 performances of Prelude in C major, Op.28, No.1 by Frédéric Chopin

Table 10.2 Correlation coefficient *r* for arousal calculated for each pair of performances of Prelude in C major, Op.28, No.1 by F. Chopin

	A. Rubinstein	E. Gilels	G. Sokolov	M. Argerich	R. Blechacz
A. Rubinstein	1.00	0.74	**0.80**	0.69	0.75
E. Gilels	0.74	1.00	0.78	0.74	**0.87**
G. Sokolov	**0.80**	0.78	1.00	0.67	**0.83**
M. Argerich	0.69	0.74	0.67	1.00	0.75
R. Blechacz	0.75	**0.87**	**0.83**	0.75	1.00

5 performances in terms of arousal, the pieces by G. Sokolov and M. Argerich were the most different ($r = 0.67$).

10.8.3   Performances and Valence over Time

Observation of the course of valence in the performances (Fig. 10.7) shows that there was a similar decrease in this value for samples 15–20 in all performances. There is a similar line shape, i.e. good correlation, between R. Blechacz and G. Sokolov (samples 5–20), and between M. Argerich and E. Gilels (samples 35–50).

Table 10.3 presents correlation coefficients for valence calculated for each pair of performances. In terms of valence distribution, G. Sokolov's performance was

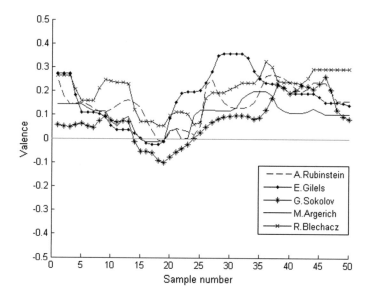

Fig. 10.7 Valence over time for 5 performances of Prelude in C major, Op.28, No.1 by Frédéric Chopin

Table 10.3 Correlation coefficient r for valence calculated for each pair of performances of Prelude in C major, Op.28, No.1 by F. Chopin

	A. Rubinstein	E. Gilels	G. Sokolov	M. Argerich	R. Blechacz
A. Rubinstein	1.00	0.41	**0.82**	0.65	0.79
E. Gilels	0.41	1.00	0.41	0.70	0.42
G. Sokolov	**0.82**	0.41	1.00	0.61	**0.81**
M. Argerich	0.65	0.70	0.61	1.00	0.70
R. Blechacz	0.79	0.42	**0.81**	0.70	1.00

similar to A. Rubinstein ($r = 0.82$) and R. Blechacz ($r = 0.81$). The performance by E. Gilels was similar to M. Argerich ($r = 0.70$), and less similar to G. Sokolov, A. Rubinstein and R. Blechacz ($r = 0.41$–0.42).

10.8.4 Performances and Arousal-Valence over Time

Another possibility of comparing performances is to take arousal and valence into account simultaneously. To compare performances described by two sequences of values (arousal and valence), these sequences should be joined. In order to join the two sequences of arousal (Eq. 10.2) and valence (Eq. 10.3) values of one performance, standard deviation and the mean of the two sequences of data should be equivalent. We decided to leave arousal values without change and convert valence values (Eq. 10.5); although we could have converted arousal and left valence without a change and this would not have affected the correlation results. During joining, the sequences of each feature are interleaved (Eq. 10.4).

Sequences for arousal and valence:

$$A = (a_1, a_2, a_3, ..., a_n) \tag{10.2}$$

$$V = (v_1, v_2, v_3, ..., v_n) \tag{10.3}$$

Result of joining arousal and valence sequence into one sequence AV:

$$AV = (a_1, vnew_1, a_2, vnew_2, a_3, vnew_3, ..., a_n, vnew_n) \tag{10.4}$$

Formula for calculating new valence sequence values:

$$vnew_n = \bar{a} + sd_a \frac{v_n - \bar{v}}{sd_v} \tag{10.5}$$

where sd_a stands for standard deviation of sequence A, sd_v—standard deviation of sequence V, $\bar{a}$—mean value of sequence A, $\bar{v}$—mean value of sequence V.

Table 10.4 presents the correlation coefficients r for joined arousal and valence, calculated for each pair of performances. We see that the most similar performances are by R. Blechacz and G. Sokolov ($r = 0.82$) and the most different by A. Rubinstein end E. Gilels ($r = 0.58$). It can be stated that in terms of arousal and valence we have two groups of performances. The first group consists of performances by R. Blechacz, G. Sokolov and A. Rubinstein, and the second group by E. Gilels and M. Argerich.

Table 10.4 Correlation coefficient r for joined arousal and valence calculated for each pair of performances of Prelude in C major, Op.28, No.1 by F. Chopin

	A. Rubinstein	E. Gilels	G. Sokolov	M. Argerich	R. Blechacz
A. Rubinstein	1.00	0.58	**0.81**	0.67	0.77
E. Gilels	0.58	1.00	0.60	0.72	0.64
G. Sokolov	**0.81**	0.60	1.00	0.64	**0.82**
M. Argerich	0.67	0.72	0.64	1.00	0.72
R. Blechacz	0.77	0.64	**0.82**	0.72	1.00

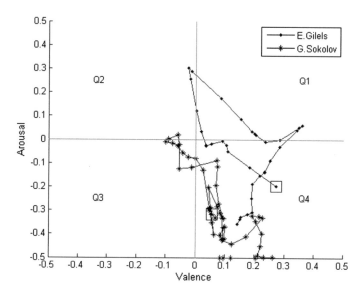

Fig. 10.8 Performances of Prelude in C major, Op.28, No.1 by Frédéric Chopin on the A-V emotion plane

10.8.5 Arousal-Valence Trajectory

Figure 10.8 presents the trajectories of two different performances (E. Gilels, G. Sokolov) of Prelude No.1 by Frédéric Chopin on the A-V emotion plane. Including all 5 performances would have obscured the illustration, therefore we decided to present two extremely different performances. A square marker on the trajectory indicates the beginning of a piece.

The trajectories illustrate how the artist moved in the 4 quarters on the Arousal-Valence emotion plane. Both performances begin and end in quarter Q4 (arousal low—valence high). The course of the middle part of these two performances varies. G. Sokolov's trajectory moves through quarter Q3 (arousal low—valence low), and E. Gilels' through quarters Q2 (arousal high—valence low) and Q1 (arousal high—valence high).

10.8.6 Scape Plot

Comparing musical performances by using correlation coefficient r calculated for the whole length of a composition is only a general analysis of similarities between recordings. In order to analyze similarities between the performances in greater detail, we used scape plotting.

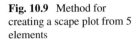

Fig. 10.9 Method for creating a scape plot from 5 elements

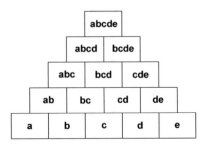

Scape plot is a plotting method that allows presenting analysis results for segments of varying lengths on one image. Their advantage is that they enable seeing the entire structure of a composition. Any type of analysis can be used during scape plot construction. In our case, analysis consisted of assessing correlations between arousal and valence values of different musical performances.

The scape plotting method was designed by Craig Sapp for structural analysis of harmony in musical scores [92]. It has also been applied to timbral analysis [98] and for visualization of the tonal content of polyphonic audio signals [28]. The scape plots were used to present similarities between tempo and loudness features extracted from recordings of the same musical composition [93, 94]. Scape plots are also used for visualizing repetitive structures of music recordings [70].

In this paper, a scape plot is used to present calculated correlations between analogous segments of the examined recordings. A comparison of sequences of emotional features (arousal and valence) in different performances of the same composition is a novel application for scape plots.

Figure 10.9 presents a method for creating a scape plot to analyze a sample composition consisting of 5 elements: a, b, c, d, e. These elements are first examined separately, and then grouped by sequential pairs: ab, bc, cd, de. Next, 3-element sequences are created: abc, bcd, cde; followed by 4-element sequences: $abcd, bcde$; and finally one sequence consisting of the entire composition: $abcde$. The obtained sequences are arranged on a plane in the form of a triangle, where at the base are the analysis results of examining the shortest sequences and at the top of the triangle are the results of analyzing the entire length of the composition. In a scape plot, the horizontal axis represents time in the composition, while the vertical axis represents the length of the analyzed sequence.

Figure 10.10 shows a sample result of creating a scale plot using arousal sequences for 5 different performances. First, a reference performance is selected for the scape plot; in this case, it was performance AAA. Then, for each cell in the scape plot, the arousal correlation between the reference performance and all other performances is calculated. Finally, the winner, i.e. the performance with the highest correlation value, is denoted using a color for each cell. Additionally, a percentage content is calculated for each winning performance.

On the provided example, reference performance AAA is most correlated with performance BBB when comparing the entire length of the composition, as

Fig. 10.10 Result of
creating a scape plot by using
arousal correlation between
the reference performance
and 4 other performances

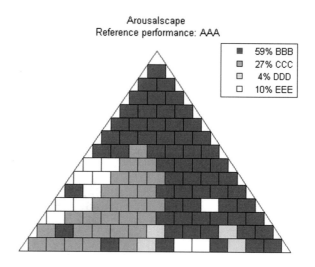

indicated by the top element of BBB in the triangle. This is additionally confirmed by the percentage of wins, i.e. the area occupied, by performance BBB (59%). Performance CCC is in second place with 27% wins. The placement of the wins for this performance, bottom left of the scape plot, indicates a similarity with the reference performance in the first half of the composition. Performances DDD and EEE showed little similarity (4 and 10% wins) with the reference performance.

10.8.7 Arousalscape

Figure 10.11 shows the Arousalscape, a scape plot generated for the arousal value sequence for 5 different performances of Prelude in C major, Op.28, No.1 by Frédéric Chopin. The performance by A. Rubinstein was selected as the reference performance.

Arousalscape illustrates which performances are the most correlated to A. Rubinstein, by different lengths of the examined sequences. In the lower levels of the triangle, it is difficult to choose the winner; but in the higher levels of the triangle, the winner is unequivocal: the performance by G. Sokolov. His area in the Arousalscape is the biggest and reaches a value of 59%. It is interesting that on the first half of Prelude, the performance by E. Gilels is the most similar to A. Rubinstein. The remaining two performances (M. Argerich, R. Blechacz) were covered by the first two winners during comparison with the reference performance (A. Rubinstein).

Fig. 10.11 Arousalscape for
Prelude in C major, Op.28,
No.1, reference
performance: A. Rubinstein

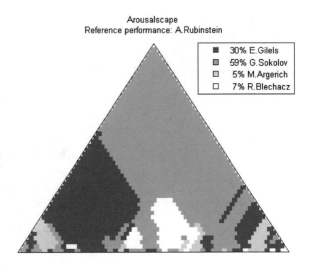

10.8.8 *Valencescape*

Figure 10.12 shows the Valencescape, a scape plot generated for the valence value
sequence for 5 different performances of Prelude in C major, Op.28, No.1 by Frédéric
Chopin. The performance by A. Rubinstein was selected as the reference performance.

The situation here is not as unequivocal as for Arousalscape for the same composition (Sect. 10.8.7). Once again the winner was the performance by G. Sokolov, but
its area on the Valencescape is smaller (49%). The triangle area is ruptured by the

Fig. 10.12 Valencescape for
Prelude in C major, Op.28,
No.1, reference
performance: A. Rubinstein

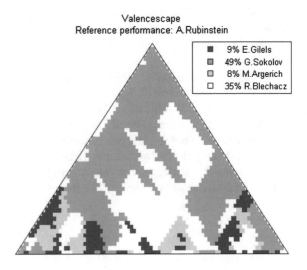

Fig. 10.13 AVscape for
Prelude in C major, Op.28,
No.1, reference
performance: A. Rubinstein

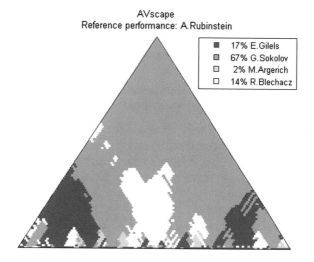

color white, which represents the performance by R. Blechacz (35% of the area). This
means that this performance was also well correlated with the reference performance
at many moments.

10.8.9   AVscape

Figure 10.13 shows the AVscape, a scape plot generated for sequences of joined
arousal and valence values for 5 different performances of Prelude in C major, Op.28,
No.1 by Frédéric Chopin. The performance by A. Rubinstein was selected as the reference performance. The definitive winner of the comparisons of correlation values
with various lengths of sequences is the performance by G. Sokolov (67% of the
area), with E. Gilels in second place (17% of the area), and R. Blechacz in third
place (14% of the area). At the beginning of the composition, the reference performance by A. Rubinstein is similar to E. Gilels (dark colored left lower part of the
triangle), and in the middle of the composition to R. Blechacz (white space in the
middle lower part of the triangle).

When comparing 5 different performances of one composition, we can create 5
AVscapes, 5 Arousalscapes, and 5 Valencescapes. Each rendition is consecutively
selected as the reference performance and the degree of similarity is visualized in
detail in comparison with the other 4 performances.

10.9   Evaluation

10.9.1   Parameters Describing the Most Similar Performances

To find the most similar performance to the reference performance, you can use several indicators that result from the construction of the obtained scape plots [94]:

- Score S_0—the most general result indicating the most similar performance. The winner is the one with the best correlation for the entire sequence, entire length of time of the composition. On the scape plot, it is the top element of the triangle.
- Score S_1—indicates the performance with the biggest area in the scape plot. The area of wins of a given performance shows its dominance at various lengths of analyzed sequences. The winner with the best correlation for the entire sequence (S_0) does not always have the largest area, or the largest number of wins on the scape plot.
- Score S_2—the next best similar performance from the scape plot, calculated after removing the winner S_1. If two performances are very similar, then one will always win and cover the wins of the second. To calculate the S_2 score, a new scape plot is generated without the winner, indicated by S_1. It shows the performance with the biggest area in the newly created scape plot.

Figure 10.14 shows the AVscape for Prelude in C minor, Op.28, No.20, reference performance: M. Argerich. Score S_0 in this case is the performance by E. Gilels—the top cell of the triangle. S_1 indicates the performance with the biggest area and that is the performance by A. Rubinstein (49% of the area).

S_2 calculations are presented in Fig. 10.15. The winner S_1 from Fig. 10.14 is removed and the performance with the biggest area (S_2) in the newly generated scape plot is the performance by E. Gilels (81% of the area).

Fig. 10.14 AVscape for
Prelude in C minor, Op.28,
No.20, reference
performance: M. Argerich

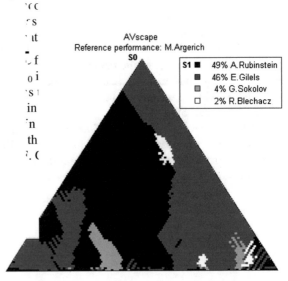

AVscape
Reference performance: M.Argerich
S0

S1 ■	49% A.Rubinstein
■	46% E.Gilels
▨	4% G.Sokolov
□	2% R.Blechacz

Fig. 10.15 AVscape with winner S_1 removed for Prelude in C minor, Op.28, No.20, reference performance: M. Argerich

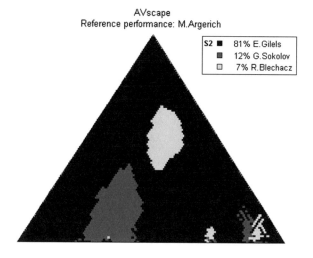

10.9.2 Ground Truth for Performing Similarity Assessments Between Performances

Determining similarities between performers is not an easy task for a human. On the one hand, specialist musical knowledge and experience are required of the evaluators. On the other hand, comparison of even short one-minute performances can cause much difficulty even for experienced musicians.

We used a similarity matrix form [5 × 5] to collect expert opinions, presented in Table 10.5, which is a symmetric matrix, with the main diagonal values equal to 10. The experts' task was to determine which performances of a given performer are the most similar and which are the least. Each expert had to determine these values in the form for each composition. Each value described the degree of similarity between performances. The filled values were subsequent natural numbers in the range of [1, 9], where 9 meant a very similar performance, and 1 very different. A value of 10 on the diagonal meant that the given performance is maximally similar. Information on the performer of a given rendition was kept from the evaluators.

Table 10.5 Form for similarity matrix between performances

	Performance 1	Performance 2	Performance 3	Performance 4	Performance 5
Performance 1	10				
Performance 2		10			
Performance 3			10		
Performance 4				10	
Performance 5					10

Table 10.6 Compositions and agreement between 3 experts' opinions

Composition	Cronbachs α	Avg. Spearman's ρ
Prelude No.1	0.76	0.51
Prelude No.5	0.73	0.48
Prelude No.18	0.77	0.54
Prelude No.20	0.68	0.42
Symphony No.3	0.95	0.86
Symphony No.5	0.84	0.64

Three music experts with a university music education participated in the experiment. It took approximately 30–40 min to compare 5 one-minute performances of one composition. This included multiple times of listening to a performance as well as finding the appropriate numerical values on the similarity matrix, which was not a trivial task.

After collecting data from the experts, the data was ranked. This way we eliminated different individual opinions on the similarity scale and maintained the sequence of degree of similarity. On different questionnaires, the most similar performances could have different values—the maximum value on a given questionnaire—but after ranking the maximum values will always be in first place.

Finally, the obtained values were averaged and rescaled to the range of [1, 9]. Thus we built a similarity matrix obtained from experts. It constituted the ground truth for similarities for a given composition and was used to compare with the matrix of similarity between performances obtained by a computer system.

To check the agreement between three experts' opinions, Cronbachs α [15] and average Spearman's ρ was calculated (Table 10.6). Spearman's ρ was calculated between each set of two individual opinions, an then the obtained values ware averaged.

The calculated parameter values confirm that the collected data from the experts are correlated. The positive avg. Spearman's ρ values (from 0.42 to 0.86) indicate a clear relation between the experts' responses, although the greatest concordance was between the opinions for Symphony No.3 (0.86). With regard to internal consistency represented by Cronbachs α, the obtained values are good and acceptable.

10.9.3 Evaluation Parameters

To assess the built system, we used a series of parameters comparing the obtained results with data obtained from music experts.

10.9.3.1 Spearman's Rank Correlation Coefficient

The first evaluation parameter we used was Spearman's rank correlation coefficient (Spearman's ρ) [13], which is the Pearson correlation coefficient between ranked variables. Before calculating correlations, variables are converted to ranks. The rank correlation coefficient ranges from 1 to -1. A positive ρ indicates a positive relationship between the two variables, while a negative ρ expresses a negative relationship.

The following formula (Eq. 10.6) is used to calculate the Spearman rank correlation between n values of two variables X_i, Y_i:

$$\rho = 1 - \frac{6 \sum_{i=1}^{n} d_i^2}{n(n^2 - 1)} \tag{10.6}$$

where $d_i = rg(X_i) - rg(Y_i)$ are the differences between ranks.

In our case, Spearman's ρ measures how much the similarity values provided by the computer system and experts have a similar rank. While calculating Spearman's ρ between the similarity matrix obtained from experts and the similarity matrix obtained from the computer system, only elements below the main diagonal from the matrix were taken into account. Matrixes are symmetric; diagonal and upper diagonal elements are irrelevant. The greater the obtained Spearman's rank correlation coefficient, the closer the system's results were to the experts' opinions.

Spearman's ρ was calculated for the results between the experts' opinions and three similarity matrices obtained from the system: arousal similarity matrix ρ_A, valence similarity matrix ρ_V, and arousal-valence similarity matrix ρ_{AV}.

10.9.3.2 Maximal Similar Number of Hits

The next parameters evaluated the concordance of the indicators on the similarity matrix obtained from the experts and the similarity matrix obtained from the system. We compared indicators of the most similar performers according to the experts and the system. First, from among the experts' opinions we found the most similar performance to the reference performance, and then checked if it was confirmed by the system. If the indicators from both sides were convergent, we had a hit. The comparisons were performed for all reference performances, and the result was a percentage of hits—*MSH* (maximal similar hits) defined in Eqs. 10.7 and 10.8.

$$MSH = \frac{\sum_{i=1}^{n} H_i}{n} \times 100\% \tag{10.7}$$

$$H_i = \begin{cases} 1 & \text{if } MS_i(EX) = MS_i(CS) \\ 0 & \text{if } MS_i(EX) \neq MS_i(CS) \end{cases} \tag{10.8}$$

where *EX* is the similarity matrix obtained from experts, *CS* is the similarity matrix obtained from the computer system, $MS_i()$ is the most similar performance to the reference performance i, and n is the number of performances.

Calculating *MSH*, we can compare the similarity matrix obtained from the experts to the similarity matrix obtained from the system on the basis of different indicators: S_0 or S_1 (Sect. 10.9.1).

To check if the searched most similar performance indicated by the experts is in the top indications by the computer system, we introduced a variant of the previous parameter—*MSH2F* (maximal similar hits 2 first). The *MSH2F* calculation checks if the most similar performance according to the experts is among the top 2 indicated by the system. In the case of comparison with the results obtained on the scape plot, the first 2 most similar performances are indicated by S_1 and S_2 ($MSH2F_{S_1 S_2}$).

10.9.3.3 Evaluation Results

The obtained results of the evaluation are presented in Table 10.7. The first columns present Spearman's ρ calculated for the results between the experts' opinions and three similarity matrices obtained from the system: arousal-valence similarity matrix ρ_{AV}, arousal similarity matrix ρ_A, and valence similarity matrix ρ_V. The positive Spearman's ρ values (avg. $\rho_{AV} = 0.53$) indicate a clear relation and accordance with the experts' opinions and the computer system's calculations. Spearman's ρ taking into account arousal and valence ρ_{AV} obtained better results on average than Spearman's ρ for arousal and valence separately ($\rho_A = 0.46$, $\rho_V = 0.47$).

MSH and *MSH2F* were calculated between the experts' opinions and the arousal-valence similarity matrix. Analyzing the indicators for the most similar performance according to the experts as well as the system, the average accuracy of the applied method was 40% when using score S_0, and 47% score S_1. However, the higher values of avg. $MSH2F_{S_0}$ and avg. $MSH2F_{S_1 S_2}$ (80 and 77%) indicate that the results provided by the experts are in the top results obtained from the system.

Table 10.7 Evaluation parameters for the analyzed compositions

Composition	ρ_{AV}	ρ_A	ρ_V	MSH_{S_0}	$MSH2F_{S_0}$	MSH_{S_1}	$MSH2F_{S_1 S_2}$
Prelude No.1	0.52	0.60	0.52	40	80	40	80
Prelude No.5	0.72	0.30	0.61	40	80	40	60
Prelude No.18	0.50	0.69	0.30	80	80	80	100
Prelude No.20	0.55	0.88	0.43	40	100	60	100
Symphony No.3	0.58	−0.12	0.74	20	80	20	80
Symphony No.5	0.34	0.41	0.22	20	60	40	40
Averages	0.53	0.46	0.47	40	80	47	77

10.10 Conclusions

In this study, we presented a comparative analysis of musical performances by using emotion tracking. Use of emotions for comparisons is a novel approach, not found in other hitherto published papers. Values of arousal and valence, predicted by regressors, were used to compare performances. We found which performances of the same composition are closer to each other and which are quite distant in terms of the shaping of arousal and valence over time. We evaluated the applied approach comparing the obtained results with the opinions of music experts. The obtained results confirm the validity of the assumption that tracking and analyzing the values of arousal and valence over time in different performances of the same composition can be used to indicate their similarities.

The presented method gives access to knowledge on the shaping of emotions by a performer, which had previously been available only to music professionals. Finding pieces with a similar emotional distribution throughout the same composition is an interesting option that further extends the capabilities of systems searching musical compositions.

References

1. Aha, D.W., Kibler, D., Albert, M.K.: Instance-based learning algorithms. Mach. Learn. **6**(1), 37–66 (1991)
2. Aljanaki, A., Wiering, F., Veltkamp, R.C.: Computational modeling of induced emotion using GEMS. In: Proceedings of the 15th International Society for Music Information Retrieval Conference, ISMIR 2014, Taipei, Taiwan, pp. 373–378 (2014)
3. Aljanaki, A., Wiering, F., Veltkamp, R.C.: Emotion based segmentation of musical audio. In: Proceedings of the 16th International Society for Music Information Retrieval Conference, ISMIR 2015, Málaga, Spain, pp. 770–776 (2015)
4. Aljanaki, A., Yang, Y.H., Soleymani, M.: Emotion in music task: lessons learned. In: Working Notes Proceedings of the MediaEval 2016 Workshop, Hilversum, The Netherlands, 20–21 October (2016)
5. Bachorik, J., Bangert, M., Loui, P., Larke, K., Berger, J., Rowe, R., Schlaug, G.: Emotion in motion: investigating the time-course of emotional judgments of musical stimuli. Music Percept. **26**, 355–364 (2009)
6. Baume, C., Fazekas, G., Barthet, M., Marston, D., Sandler, M.: Selection of audio features for music emotion recognition using production music. In: Audio Engineering Society Conference: 53rd International Conference: Semantic Audio (2014)
7. Bello, J.P., Duxbury, C., Davies, M., Sandler, M.: On the use of phase and energy for musical onset detection in the complex domain. IEEE Sig. Process. Lett. **11**(6), 553–556 (2004)
8. Bogdanov, D., Wack, N., Gómez, E., Gulati, S., Herrera, P., Mayor, O., Roma, G., Salamon, J., Zapata, J., Serra, X.: ESSENTIA: an audio analysis library for music information retrieval. In: Proceedings of the 14th International Society for Music Information Retrieval Conference, Curitiba, Brazil, pp. 493–498 (2013)
9. Bresin, R., Friberg, A.: Emotional coloring of computer-controlled music performances. Comput. Music J. **24**(4), 44–63 (2000)
10. Brossier, P., Bello, J.P., Plumbley, M.D.: Fast labelling of notes in music signals. In: Proceedings of the ISMIR 2004, 5th International Conference on Music Information Retrieval, Barcelona, Spain, 10–14 October (2004)
11. Cabrera, D.: Psysound: a computer program for psychoacoustical analysis. In: Proceedings of the Australian Acoustical Society Conference, pp. 47–54 (1999)
12. Cannam, C., Landone, C., Sandler, M.: Sonic visualiser: an open source application for viewing, analysing, and annotating music audio files. In: Proceedings of the ACM Multimedia 2010 International Conference, Firenze, Italy, pp. 1467–1468 (2010)
13. Chen, P.Y., Popovich, P.M.: Correlation: parametric and nonparametric measures. Sage, Thousand Oaks (Calif.) (2002)
14. Chen, Y.A., Yang, Y.H., Wang, J.C., Chen, H.: The amg1608 dataset for music emotion recognition. In: 2015 IEEE International Conference on Acoustics, Speech and Signal Processing (ICASSP), pp. 693–697 (2015)

© Springer International Publishing AG 2018
J. Grekow, *From Content-Based Music Emotion Recognition to Emotion Maps of Musical Pieces*, Studies in Computational Intelligence 747,
https://doi.org/10.1007/978-3-319-70609-2

15. Cronbach, L.J.: Coefficient alpha and the internal structure of tests. Psychometrika **16**(3), 297–334 (1951)
16. Deng, J.J., Leung, C.H.: Dynamic time warping for music retrieval using time series modeling of musical emotions. IEEE Trans. Affect. Comput. **6**(2), 137–151 (2015)
17. DiPaola, S., Arya, A.: Emotional remapping of music to facial animation. In: ACM Siggraph 2006 Video Game Symposium Proceedings (2006)
18. Dixon, S., Widmer, G.: MATCH: A music alignment tool chest. In: Proceedings of the ISMIR 2005, 6th International Conference on Music Information Retrieval, London, UK, 11–15 September 2005, pp. 492–497 (2005)
19. Eerola, T., Toiviainen, P.: MIDI Toolbox: MATLAB Tools for Music Research. University of Jyväskylä, Jyväskylä (2004)
20. Eerola, T., Vuoskoski, J.K.: A comparison of the discrete and dimensional models of emotion in music. Psychol. Music **39**(1), 18–49 (2011)
21. Ekman, P.: An argument for basic emotions. Cogn. Emot. **6**(3), 169–200 (1992)
22. Ekman, P.: Basic emotions. In: Dalgleish, T., Powers, M.J. (eds.) Handbook of Cognition and Emotion, pp. 4–5. Wiley, Hoboken (1999)
23. Farnsworth, P.R.: A study of the hevner adjective list. J. Aesthet. Art Crit. **13**(1), 97–103 (1954)
24. Frank, E., Witten, I.H.: Generating accurate rule sets without global optimization. In: Proceedings of the Fifteenth International Conference on Machine Learning, pp. 144–151. Morgan Kaufmann Publishers Inc., San Francisco (1998)
25. Gabrielsson, A.: Emotion perceived and emotion felt: same or different? Musicae Scient. **5**(1 suppl), 123–147 (2002)
26. Goebl, W., Pampalk, E., Widmer, G.: Exploring expressive performance trajectories: six famous pianists play six chopin pieces. In: Proceedings of the 8th International Conference on Music Perception and Cognition (ICMPC'8), Evanston, IL, USA, pp. 505–509 (2004)
27. Gómez, E.: Tonal description of music audio signals. Ph.D. thesis, Universitat Pompeu Fabra (2006)
28. Gómez, E., Bonada, J.: Tonality visualization of polyphonic audio. In: Proceedings of the International Computer Music Conference, Barcelona, Spain (2005)
29. Good, M.: Musicxml for notation and analysis. Virtual score Represent. Retr. Restor. **12**, 113–124 (2001)
30. Grekow, J.: Broadening musical perception by akwets technique visualisation. In: Proceedings of the 9th International Conference on Music Perception and Cognition, ICMPC9. Bononia University Press, Bologna (2006)
31. Grekow, J.: Emotion based music visualization system. In: M. Kryszkiewicz, H. Rybinski, A. Skowron, Z.W. Raś (eds.) Proceedings of the Foundations of Intelligent Systems: 19th International Symposium, ISMIS 2011, Warsaw, Poland, 28–30 June 2011, pp. 523–532. Springer, Berlin (2011)
32. Grekow, J.: Mood tracking of musical compositions. In: L. Chen, A. Felfernig, J. Liu, Z.W. Raś (eds.) Proceedings of the Foundations of Intelligent Systems: 20th International Symposium, ISMIS 2012, Macau, China, 4–7 December 2012. Springer, Berlin (2012)
33. Grekow, J.: Method of transforming music into 3D figures. Prz. Elektrotech. **89**(11), 327–330 (2013)
34. Grekow, J.: Mood tracking of radio station broadcasts. In: Andreasen, T., Christiansen, H., Cubero, J.C., Raś, Z.W. (eds.) Foundations of Intelligent Systems. Lecture Notes in Computer Science, vol. 8502, pp. 184–193. Springer International Publishing, Cham (2014)
35. Grekow, J.: Audio features dedicated to the detection of four basic emotions. In: Proceedings of the Computer Information Systems and Industrial Management: 14th IFIP TC 8 International Conference, CISIM 2015, Warsaw, Poland, 24–26 September 2015, pp. 583–591. Springer International Publishing, Cham (2015)
36. Grekow, J.: Emotion detection using feature extraction tools. In: F. Esposito, O. Pivert, M.S. Hacid, Z.W. Rás, S. Ferilli (eds.) Proceedings of the Foundations of Intelligent Systems: 22nd International Symposium, ISMIS 2015, Lyon, France, 21–23 October 2015, pp. 267–272. Springer International Publishing, Cham (2015)

37. Grekow, J.: Method of transforming music into 4D figures. Prz. Elektrotech. **91**(4), 159–162 (2015)
38. Grekow, J.: Music emotion maps in arousal-valence space. In: Proceedings of the Computer Information Systems and Industrial Management: 15th IFIP TC8 International Conference, CISIM 2016, Vilnius, Lithuania, pp. 697–706. Springer International Publishing, Cham (2016)
39. Grekow, J., Raś, Z.: Detecting emotions in classical music from midi files. In: Proceedings of the 18th International Conference on Foundations of Intelligent Systems, ISMIS'09, pp. 261–270. Springer, Berlin (2012)
40. Grekow, J., Raś, Z.W.: Emotion based midi files retrieval system. In: Raś, Z.W., Wieczorkowska, A. (eds.) Advances in Music Information Retrieval, pp. 261–284. Springer, Berlin (2010)
41. Hall, M., Frank, E., Holmes, G., Pfahringer, B., Reutemann, P., Witten, I.H.: The weka data mining software: an update. SIGKDD Explor. Newsl. **11**(1), 10–18 (2009)
42. Hevner, K.: Experimental studies of the elements of expression in music. Am. J. Psychol. **48**(2), 246–268 (1936)
43. Hu, X., Downie, J.S.: Exploring mood metadata: relationships with genre, artist and usage metadata. In: Proceedings of the 8th International Conference on Music Information Retrieval, ISMIR 2007, Vienna, Austria, 23–27 September 2007, pp. 67–72 (2007)
44. Hu, X., Downie, J.S., Laurier, C., Bay, M., Ehmann, A.F.: The 2007 MIREX audio mood classification task: lessons learned. In: ISMIR 2008, 9th International Conference on Music Information Retrieval, Drexel University, Philadelphia, PA, USA, 14–18 September 2008, pp. 462–467 (2008)
45. Huron, D.: Music information processing using the humdrum toolkit: concepts, examples, and lessons. Comput. Music J. **26**(2), 11–26 (2002)
46. Imbrasaite, V., Baltrusaitis, T., Robinson, P.: Emotion tracking in music using continuous conditional random fields and relative feature representation. In: 2013 IEEE International Conference on Multimedia and Expo Workshops, San Jose, CA, USA, pp. 1–6 (2013)
47. Johnson-Laird, P.N., Oatley, K.: The language of emotions: an analysis of a semantic field. Cogn. Emot. **3**(2), 81–123 (1989)
48. Juslin, P.N., Laukka, P.: Expression, perception, and induction of musical emotions: a review and a questionnaire study of everyday listening. J. New Music Res. **33**(3), 217–238 (2004)
49. Kim, Y.E., Schmidt, E.M., Migneco, R., Morton, B.G., Richardson, P., Scott, J.J., Speck, J.A., Turnbull, D.: State of the art report: music emotion recognition: a state of the art review. In: Proceedings of the 11th International Society for Music Information Retrieval Conference, ISMIR 2010, Utrecht, Netherlands, pp. 255–266 (2010)
50. Kohavi, R., John, G.H.: Wrappers for feature subset selection. Artif. Intell. **97**(1–2), 273–324 (1997)
51. Korhonen, M.D., Clausi, D.A., Jernigan, M.E.: Modeling emotional content of music using system identification. Trans. Sys. Man Cyber. Part B **36**(3), 588–599 (2005)
52. Kostek, B., Plewa, M.: Rough sets applied to mood of music recognition. In: 2016 Federated Conference on Computer Science and Information Systems (FedCSIS), pp. 71–78 (2016)
53. Lartillot, O., Toiviainen, P.: MIR in matlab (II): a toolbox for musical feature extraction from audio. In: Proceedings of the 8th International Conference on Music Information Retrieval, ISMIR 2007, Vienna, Austria, pp. 127–130 (2007)
54. Laurier, C.: Automatic classification of musical mood by content-based analysis. Ph.D. thesis, Universitat Pompeu Fabra (2011)
55. Laurier, C., Herrera, P.: Audio music mood classification using support vector machine. In: International Society for Music Information Research Conference (ISMIR) (2007)
56. Laurier, C., Herrera, P.: Mood cloud: A real-time music mood visualization tool. In: CMMR, Computer Music Modeling and Retrieval. Copenhagen (2008)
57. Li, T., Ogihara, M.: Detecting emotion in music. In: Proceedings of the ISMIR 2003, 4th International Conference on Music Information Retrieval, Baltimore, Maryland, USA, 27–30 October (2003)

58. Lidy, T., Rauber, A.: Visually profiling radio stations. In: Proceedings of the ISMIR 2006, 7th International Conference on Music Information Retrieval, Victoria, Canada, 8–12 October 2006, pp. 186–191 (2006)
59. Liem, C.C.S., Hanjalic, A.: Comparative analysis of orchestral performance recordings: an image-based approach. In: Proceedings of the 16th International Society for Music Information Retrieval Conference, ISMIR 2015, Málaga, Spain, pp. 302–308 (2015)
60. Lin, Y., Chen, X., Yang, D.: Exploration of music emotion recognition based on MIDI. In: Proceedings of the 14th International Society for Music Information Retrieval Conference, ISMIR 2013, Curitiba, Brazil, pp. 221–226 (2013)
61. Livingstone, S.R., Brown, A.R.: Dynamic response: real-time adaptation for music emotion. In: Proceedings of the Second Australasian Conference on Interactive Entertainment, IE 2005, pp. 105–111. Creativity and Cognition Studios Press, Sydney (2005)
62. Logan, B.: Mel frequency cepstral coefficients for music modeling. In: Proceedings of the ISMIR 2000, 1st International Symposium on Music Information Retrieval, Plymouth, Massachusetts, USA, 23–25 October (2000)
63. Lu, L., Liu, D., Zhang, H.J.: Automatic mood detection and tracking of music audio signals. Trans. Audio Speech Lang. Proc. **14**(1), 5–18 (2006)
64. MacDorman, K.F., Ough, S., Ho, C.C.: Automatic emotion prediction of song excerpts: index construction, algorithm design, and empirical comparison. J. New Music Res., 281–299 (2007)
65. McEnnis, D., McKay, C., Fujinaga, I., Depalle, P.: jAudio: an feature extraction library. In: ISMIR 2005, 6th International Conference on Music Information Retrieval, London, UK, pp. 600–603 (2005)
66. McKay, C., Fujinaga, I.: jSymbolic: a feature extractor for midi files. In: ICMC (2006)
67. Mehrabian, A., Russell, J.A.: An Approach to Environmental Psychology. MIT Press, Cambridge (1974)
68. Mohammad, S.: From once upon a time to happily ever after: tracking emotions in novels and fairy tales. In: Proceedings of the 5th ACL-HLT Workshop on Language Technology for Cultural Heritage, Social Sciences, and Humanities, LaTeCH 2011, pp. 105–114. Association for Computational Linguistics, Stroudsburg (2011)
69. Moriguchi, N., Wada, E., Miura, M.: Emotion control system for midi excerpts: Mor2art. In: Proceedings of the 3rd International Conference on Fun and Games, Fun and Games 2010, pp. 58–65. ACM, New York (2010)
70. Müller, M., Jiang, N.: A scape plot representation for visualizing repetitive structures of music recordings. In: Proceedings of the 13th International Society for Music Information Retrieval Conference, ISMIR 2012, Mosteiro S.Bento Da Vitória, Porto, Portugal, 8–12 October 2012, pp. 97–102 (2012)
71. Wang, M.Y., Zhang, N.Y., Zhu, H.: User-adaptive music emotion recognition. In: Proceedings of the IEEE International Conference on Signal Processing, pp. 1352–1355 (2004)
72. Pampalk, E., Rauber, A., Merkl, D.: Content-based organization and visualization of music archives. In: Proceedings of the Tenth ACM International Conference on Multimedia, MULTIMEDIA 2002, pp. 570–579. ACM (2002)
73. Panda, R., Rocha, B., Paiva, R.P.: Music emotion recognition with standard and melodic audio features. Appl. Artif. Intell. **29**(4), 313–334 (2015)
74. Peeters, G.: A large set of audio feature for sound description (similarity and classification) in the cuidado project. Technical report, Ircam, Analysis/Synthesis Team, 1 pl. Igor Stravinsky, 75004 Paris, France (2004)
75. Peng, C.K., Buldyrev, S.V., Havlin, S., Simons, M., Stanley, H.E., Goldberger, A.L.: Mosaic organization of dna nucleotides. Phys. Rev. E **49**, 1685–1689 (1994)
76. Peretz, I.: Brain specialization for music. Ann. New York Acad. Sci. **930**(1), 153–165 (2001)
77. Pesek, M., Godec, P., Poredos, M., Strle, G., Guna, J., Stojmenova, E., Pogacnik, M., Marolt, M.: Introducing a dataset of emotional and color responses to music. In: Proceedings of the 15th International Society for Music Information Retrieval Conference, ISMIR 2014, Taipei, Taiwan, 27–31 October 2014, pp. 355–360 (2014)

78. Platt, J.: Fast training of support vector machines using sequential minimal optimization. In: Advances in Kernel Methods–Support Vector Learning. MIT Press, Cambridge (1998)
79. Platt, J.: Fast training of support vector machines using sequential minimal optimization. In: Schölkopf, B., Burges, C., Smola, A. (eds.) Advances in Kernel Methods–Support Vector Learning, pp. 42–65. MIT Press, Cambridge (1998)
80. Plewa, M., Kostek, B.: Multidimensional scaling analysis applied to music mood recognition. In: Audio Engineering Society Convention 134 (2013)
81. Pratt, C.C.: Music as the language of emotion. U.S. Govt. Print. Off, The Library of Congress, Washington (1950)
82. Quinlan, J.R.: C4.5: Programs for Machine Learning. Morgan Kaufmann Publishers Inc., San Francisco (1993)
83. Quinlan, R.J.: Learning with continuous classes. In: 5th Australian Joint Conference on Artificial Intelligence, pp. 343–348. World Scientific, Singapore (1992)
84. Rabiner, L., Juang, B.H.: Fundamentals of Speech Recognition. Prentice-Hall Inc, Upper Saddle River (1993)
85. Raś, Z.W., Wieczorkowska, A. (eds.): Advances in Music Information Retrieval. Studies in Computational Intelligence, vol. 274. Springer, Berlin (2010)
86. Rizk, Y., Safieddine, M., Matchoulian, D., Awad, M.: Face2mus: a facial emotion based internet radio tuner application. In: MELECON 2014 - 2014 17th IEEE Mediterranean Electrotechnical Conference, pp. 257–261 (2014)
87. Rothstein, J.: MIDI: A comprehensive introduction, vol. 7. AR Editions, Inc., Madison (1995)
88. Russell, J.A.: A circumplex model of affect. J. Personal. Soc. Psychol. **39**(6), 1161–1178 (1980)
89. Saari, P., Eerola, T., Fazekas, G., Barthet, M., Lartillot, O., Sandler, M.B.: The role of audio and tags in music mood prediction: a study using semantic layer projection. In: Proceedings of the 14th International Society for Music Information Retrieval Conference, ISMIR 2013, Curitiba, Brazil, pp. 201–206 (2013)
90. Saari, P., Eerola, T., Lartillot, O.: Generalizability and simplicity as criteria in feature selection: application to mood classification in music. IEEE Trans. Audio Speech Lang. Process. **19**(6), 1802–1812 (2011)
91. Salembier, P., Sikora, T.: Introduction to MPEG-7: Multimedia Content Description Interface. Wiley, New York (2002)
92. Sapp, C.S.: Harmonic visualizations of tonal music. In: Proceedings of the 2001 International Computer Music Conference, ICMC 2001, Havana, Cuba, 17–22 September 2001 (2001)
93. Sapp, C.S.: Comparative analysis of multiple musical performances. In: Proceedings of the 8th International Conference on Music Information Retrieval, ISMIR 2007, Vienna, Austria, pp. 497–500 (2007)
94. Sapp, C.S.: Hybrid numeric/rank similarity metrics for musical performance analysis. In: ISMIR 2008, 9th International Conference on Music Information Retrieval, Drexel University, Philadelphia, PA, USA, 14–18 September 2008, pp. 501–506 (2008)
95. Schmidt, E.M., Kim, Y.E.: Modeling musical emotion dynamics with conditional random fields. In: Proceedings of the 2011 International Society for Music Information Retrieval Conference, pp. 777–782 (2011)
96. Schmidt, E.M., Turnbull, D., Kim, Y.E.: Feature selection for content-based, time-varying musical emotion regression. In: Proceedings of the International Conference on Multimedia Information Retrieval, MIR 2010, pp. 267–274. ACM, New York (2010)
97. Schubert, E.: Update of the hevner adjective checklist. Percept. Mot. Skills **96**(4), 1117–1122 (2003)
98. Segnini, R., Sapp, C.: Scoregram: Displaying Gross Timbre Information from a Score, pp. 54–59. Springer, Berlin (2006)
99. Smola, A.J., Schölkopf, B.: A tutorial on support vector regression. Stat. Comput. **14**(3), 199–222 (2004)
100. Song, Y., Dixon, S., Pearce, M.: Evaluation of musical features for emotion classification. In: Proceedings of the 13th International Society for Music Information Retrieval Conference, ISMIR 2012, Mosteiro S.Bento Da Vitória, Porto, Portugal, pp. 523–528 (2012)

101. Streich, S., Herrera, P.: Detrended fluctuation analysis of music signals danceability estimation and further semantic characterization. In: AES 118th Convention (2005)
102. Temperley, D.: The Cognition of Basic Musical Structures. MIT Press, New York (2004)
103. Thayer, R.E.: The Biopsychology of Mood and Arousal. Oxford University Press, Cambridge (1989)
104. Trohidis, K., Tsoumakas, G., Kalliris, G., Vlahavas, I.P.: Multi-label classification of music into emotions. In: ISMIR 2008, 9th International Conference on Music Information Retrieval, Drexel University, Philadelphia, PA, USA, pp. 325–330 (2008)
105. Turnbull, D., Barrington, L., Torres, D., Lanckriet, G.: Semantic annotation and retrieval of music and sound effects. IEEE Trans. Audio Speech Lang. Process. **16**(2), 467–476 (2008)
106. Tzanetakis, G., Cook, P.: Marsyas: a framework for audio analysis. Org. Sound **4**(3), 169–175 (2000)
107. Tzanetakis, G., Cook, P.: Musical genre classification of audio signals. IEEE Trans. Speech Audio Process. **10**(5) (2002)
108. Vuoskoski, J.K.: Emotions Represented and Induced by Music: The Role of Individual Differences. University of Jyväskylä, Jyväskylä (2012)
109. Wang, X., Wu, Y., Chen, X., Yang, D.: Enhance popular music emotion regression by importing structure information. In: 2013 Asia-Pacific Signal and Information Processing Association Annual Summit and Conference, pp. 1–4 (2013)
110. Wang, Y., Witten, I.H.: Induction of model trees for predicting continuous classes. In: Poster papers of the 9th European Conference on Machine Learning. Springer, Berlin (1997)
111. Watt, R., Ash, R.: A psychological investigation of meaning in music. Musicae Sci. **2**(1), 33–53 (1998)
112. Widmer, G., Goebl, W.: Computational models of expressive music performance: the state of the art. J. New Music Res. **33**(3), 203–216 (2004)
113. Wieczorkowska, A., Synak, P., Raś, Z.W.: Multi-label classification of emotions in music. In: Intelligent Information Processing and Web Mining: Proceedings of the International IIS: IIPWM'06 Conference held in Ustroń, Poland, 19–22 June 2006, pp. 307–315. Springer, Berlin (2006)
114. Witten, I.H., Frank, E.: Data Mining: Practical Machine Learning Tools and Techniques. Morgan Kaufmann, San Francisco (2005)
115. Xiao, Z., Dellandrea, E., Dou, W., Chen, L.: What Is the best segment duration for music mood analysis? In: 2008 International Workshop on Content-Based Multimedia Indexing, pp. 17–24 (2008)
116. Xu, J., Li, X., Hao, Y., Yang, G.: Source separation improves music emotion recognition. In: Proceedings of International Conference on Multimedia Retrieval, ICMR 2014, pp. 423–426. ACM, New York (2014)
117. Xu, L., Yan, P., Chang, T.: Best first strategy for feature selection. In: Proceedings of the 9th International Conference on Pattern Recognition, vol. 2, pp. 706–708 (1988)
118. Yang, Y.H., Chen, H.H.: Machine recognition of music emotion: a review. ACM Trans. Intell. Syst. Technol. **3**(3), 40:1–40:30 (2012)
119. Yang, Y.H., Lin, Y.C., Su, Y.F., Chen, H.H.: A regression approach to music emotion recognition. Trans. Audio Speech Lang. Proc. **16**(2), 448–457 (2008)
120. Yeh, J.H., Pao, T.L., Pai, C.Y., Cheng, Y.M.: Tracking and Visualizing the Changes of Mandarin Emotional Expression, pp. 978–984. Springer, Berlin (2008)
121. Zwicker, E.: Subdivision of the audible frequency range into critical bands (frequenzgruppen). J. Acoust. Soc. Am. **33**(2), 248 (1961)

Index

© Springer International Publishing AG 2018
J. Grekow, *From Content-Based Music Emotion Recognition to Emotion
Maps of Musical Pieces*, Studies in Computational Intelligence 747,
https://doi.org/10.1007/978-3-319-70609-2

Printed in the United States
By Bookmasters